T0321372

# CHEMISTRY AND FATE OF
# ORGANOPHOSPHORUS COMPOUNDS

CURRENT TOPICS IN ENVIRONMENTAL AND
TOXICOLOGICAL CHEMISTRY
Edited by R. W. Frei and O. Hutzinger

ISSN 0275-2581
This book is part of a series. The publisher will accept continuation orders which may
be cancelled at any time and which provide for automatic billing and shipping of
each title in the series upon publication. Please write for details.

# CHEMISTRY AND FATE
# OF
# ORGANOPHOSPHORUS
# COMPOUNDS

*Edited by*
### ERNEST MERIAN
*International Association of*
*Environmental Analytical Chemistry, Therwil, Switzerland*

### ROLAND W. FREI
*Department of Analytical Chemistry,*
*The Free University, Amsterdam, The Netherlands*

### J. F. LAWERENCE
*Food Directorate, Health Protection Research,*
*Ottawa, Ontario, K1A 0, Canada*

### UDO A. Th. BRINKMAN
*Department of Analytical Chemistry,*
*The Free University, Amsterdam, The Netherlands*

GORDON AND BREACH SCIENCE PUBLISHERS
New York    London    Paris    Montreux    Tokyo    Melbourne

© 1987 by Gordon and Breach Science Publishers S.A., Post Office Box 161, 1820 Montreux 2, Switzerland. All rights reserved.

Gordon and Breach Science Publishers

Post Office Box 786
Cooper Station
New York, New York 10276
United States of America

14-9 Okubo 3-chome,
Shinjuku-ku,
Tokyo 160
Japan

Post Office Box 197
London WC2E 9PX
England

Camberwell Business Center
Private Bag 30
Camberwell, Victoria 3124
Australia

58, rue Lhomond
75005 Paris
France

The articles published in this book first appeared in *International Journal of Environmental Analytical Chemistry,* Volume 20, Numbers 2 and 4, Volume 21, Numbers 2 and 4, Volume 22, Numbers 2 and 4 and Volume 23, Numbers 2 and 3.

© 1987 Gordon and Breach, Science Publishers, Inc.

**Library of Congress Cataloging-in-Publication Data**

Chemistry and fate of organophosphorus compounds.
  (Current topics in environmental and toxicological chemistry; v. 12)
  Includes index.
  1. Organophosphorous compounds—Analysis.
2. Environmental chemistry. I. Merian, E. II. Series.
QD412.P1C44  1987    547'.07    87-8566
ISBN 2-88124-215-4

No part of this book may be reproduced or utilized in any form or by any means, electronic or mechanical, including photocopying and recording, or by any information storage or retrieval system, without permission in writing from the publishers. Printed in Great Britain by Bell and Bain Ltd., Glasgow.

# Contents

# Preface

This book comprises a selection of contributions given at the Workshop on "Chemistry and Fate of Organophosphorus Compounds" held at the Free University, Amsterdam in June 1986. The Papers presented in this volume have already been published as regular papers in the *International Journal of Environmental Analytical Chemistry* and the journal *Toxicological and Environmental Chemistry*: however, a few contributions mentioned below have not been published in these journals and so do not appear in this volume.

# Report from the Workshop

E. MERIAN

*International Association of Environmental Analytical Chemistry,
Therwil, Switzerland*

This Workshop was organized by the International Association of
Environmental Analytical Chemistry. Under the Chairmanship of
Professor Brinkman, Professor Frei, Dr. Merian and Dr. Verweij,
chemists (environmental chemists, persistence control, production),
analytical chemists, and biologists (who use analytical chemistry and
data to look at metabolisms and kinetics) discussed in an inter-
disciplinary way relevant scientific problems connected with
organophosphorus compounds. These compounds are important as
natural compounds, as pesticides and as potential chemical weapons.
Stereo-isomers may have quite different activities.

## 1. ENVIRONMENTAL CHEMISTRY, ECOSYSTEMS, DETECTION DEVICES

H. W. Houx (Institute for Pesticide Research, NL-6709 PG
Wageningen) discussed the behaviour of parathion in an aqueous
micro-ecosystem. He studied especially small polluted ditches and
leaks under Dutch conditions (many green-houses below sea level).
Realistic models of micro-ecosystems have to be handled in an easy

way (after acclimation and incubation), and aerobic and anaerobic processes must be combined. Sediments containing pesticides and other agricultural wastes are brought into compact aquaria. Sediments (after centrifugation and extraction) and water (after absorption and concentration) are analysed with HPLC. When looking at parameters, it was found that some experimental parameters have practically no influence. Amino-parathion is absorbed quite strongly to sediments, and is then difficult to extract. The author worked especially with $^{14}$C-radioactivity detectors. Wolfgang Schwack (Institute for Pharmacy and Food Chemistry, University of D-8700 Würzburg) reported on photoinduced addition of pesticides to biomolecules (parathion as an example). Lipophilic insecticides— such as parathion—are only distributed in the outer layers of plants, and thus the author studied decomposition by UV-radiation in the presence of unsaturated compounds (alkenes, terpenes, sterols, or as a model substance cyclohexene) of plant cuticles, and in presence of methyl oleate. The nitro-group is preferably reduced, but the central first nitroso product could not be isolated. Main products are N,O-dialkenyl hydroxylamino, cis/trans azo, and azoxy derivatives of parathion. The orange-red coloured azo dyes were not only identified in the laboratory, but also in apple peels. Niklaus Graber (Ciba-Geigy AG, CH-4002 Basle) considered ambient monitoring in production plants, where harmful compounds are either used as a reagent in synthesis or where intermediates or end-products might be harmful for the people working there. Of course it is always better to localize and to eliminate a problem before monitoring. Ecological processes are often also more economic. Processes, people, ambient air, emissions, imissions and waste water obviously have to be controlled. As an example, the author discussed measures in the production of the insecticide Diazinon (b.p. 84°C, toxic via skin, swallowing and inhalation) from PSE (chlorothiophosphoric acid diethylester, b.p. 49°C, acutely toxic, inhibits cholinesterase and fermentase). A special GC-prototype was developed, and the equipment (constantly controlled for proper functioning) now consists of the system probe→absorption tube→column→detector→recorder resp. alarm (when concentrations reach and pass calibrated concentrations). However, from the discussion it turned out that hazardous impurities are not yet considered.

Albert Verweij and A. W. Barendsz (Prins Maurits Laboratorium,

TNO, NL-2280 AA Rijswijk) compared various means for the detection of organophosphorus compounds (which consists in chemical war mostly in monitoring vapours). Colorimetric reactions (e.g. with the molybdenum blue method, or with the Schoenemann reaction, benzidine reaction and oxidation to an azo dyestuff), electrochemical methods, and—in future—microsensors (using transduction into electrical signals) are of primary importance for identification of Tabun, Sarin, Soman and VX, and their metabolites. For electrochemical detection the USA-ICAD alarm system is applicable. Photometric detection is less selective. Enzymatic DCI-detection (detection device, chemical agent) makes use of a 1,4-benzoquinone-monoimine to get a blue dyestuff. Optical wave guides, semiconductors and piezoelectric devices were described for chemical microsensors. The SAW chemosensor can transform detection into an acoustic signal (unfortunately with some delay). The author discussed conflicting sensor characteristics: chemi-sorption is, for instance, better regarding selectivity, whereas adsorption has merits for reversibility. Coordinating agents allow some compromises. In the general discussion it was pointed out that some older analytical measurements may not have been adequately defined and are thus not really convincing. One has primarily to look at biochemical effects, and thus at analytical chemistry relevant to biological studies. In environmental chemistry, impurities may play an important role, and perhaps controls with thin-layer chromatography for overview should be recommended before application of sophisticated methods. Recovery of standards, and fate studies of bound metabolites give further information. In summary, multi-disciplinary understanding must be improved. For analytical quantitation simpler and quicker methods are looked for, which are still adequately precise.

## 2. CHROMATOGRAPHIC ANALYTICAL CHEMISTRY (ALSO FOR KINETIC STUDIES)

James F. Lawrence (Bureau of Chemical Safety, Ottawa, Ontario K1A 0L2, Canada) presented an overview on methodology for the determination of organophosphorus pesticide residues in food. In Canada 38 OPC's are registered, but it must be considered that food with other residues is also imported, mainly from the U.S.A. and Mexico. Residue limits depend on toxicology and on food intake (they must be smaller for potatoes than for spices). For 12 OPC's concentrations must always be below 0.1 ppm. The author described the monitoring of fruits, vegetables, and meat: acetone extraction, $MeCl_2$ + hexane partition, evaporation to dryness, dissolving in hexane or acetone, florisil column clean-up or gel permeation clean-up (on a Biobeds SX-3 column) are normal procedures (fats may be dissolved directly). Recovery of Fenthion and of Ethion in vegetable oils causes some problems. Derivatisation of decomposition products may be useful. For screening (e.g. for the detection of ng's in beef liver extracts) silica gel thin-layer chromatography is recommended. Concerning confirmation with detectors, NPD's (nitrogen phosphate detectors) detect more substances, whereas FPD's (flame photometric detectors) select more specifically. A. P. Woodbridge (Shell Research Centre, Sittingbourne, Kent ME9 8AG, U.K.) discussed specific trace analysis methods for vinyl phosphate insecticides. For instance for the detection of Chlorfenvinphos, Dichlorvinphos, and Tetrachlorvinphos in waters, orange peels, cotton seed, carrots, tea, sheep fat and muscle, peat soil, sandy loam soil, air, etc. the developed technique consists in normal extraction with acetone/hexane, clean-up, quantitative GC analysis, and with NPD detection (which is rather easier to apply than ECD; with ECD it is possible to detect 0.005 ng, with FPD 0.1 ng). Monocrotophos, Dicrotophos and Mevinphos are better extracted with chloroform before GC/FPD identification (detection limits about 0.1 ng). In some cases dichloromethane or acetonitrile extraction can also be recommended. The author also compared HPLC (to detect enzyme inhibition) and gel permeation (for clean-up) with gas chromatography, and mentioned that immuno assay techniques are promising. Albert Verweij (Prins Maurits Laboratorium TNO, NL-2280 AA Rijswijk) spoke on special techniques to analyse a number of enantiomeric and diastereoisomeric organophosphorus compounds, using capillary columns coated with chiral stationary phases (temperature influence resolution: at 80°C peaks are not very broad). He referred also to the differentiation of neurotoxins of the R-O-(Y)-PO-X series by H. P. Benshop et al. (Anal. Biochem., **151**, 242, 1985). With polar (Triton X-305), apolar (DC-550), Carbowax 20 M and SE-30 packed columns it is only possible to separate C(+)P(+) and C(-)P(-)

from $C(+)P(-)$ and $C(-)P(+)$. The chirasyl stationary phase Chirasil-Val (H. Frank *et al.*, *Angew. Chem. Int. Ed. Engl.* **17,** 363, 1978) differentiates better, especially when combined with Carbowax 20 M. There are Type I (somewhat inferior) and Type II of Chirasil-Val, and additional deactivation gives improved results. The author separated even deuteriated compounds. An alternative are chiral nickel complexes (e.g. Ni(II)-bis-(heptafluorobutyryl-1-R/S-camphorate described by V. Schurig *et al.*, *J. Am. Chem. Soc.* **104,** 7573, 1982). With varying elution order it is possible to separate Tabun and Soman on OV-101/Ni(HFB-1R-Cam)$_2$ columns.

Mrs Jenny Gluckman (Dept. of Analytical Chemistry, Free University, NL-1081 HV Amsterdam) further developed organophosphorus analysis by using on-line flame and thermionic detectors in narrow bore liquid chromatography. She insisted again on the importance of detecting those compounds which are really relevant. Especially with the very efficient small columns (having a low volumetric flow rate, such as slurry-packed columns, packed capillary columns, and open-tubular capillary columns) selection of best detectors to handle sulfur and phosphorus compounds gets critical. The right specific geometry of the equipment is necessary to detect and record for instance $S_2^*$ and HP*O fragments. Flame photometric detectors have a sensitivity down to about 70 pg phosphorus/sec. Thermionic detectors collect electrons developing from produced Rb* and can thus reach detection limits of 0.2–0.5 pg phosphorus/sec. Response is linear depending on the mass of phosphorus. The author optimized interfaces, and explained practical examples for onion (spiked with 3.6 ppm Diazinon), cabbage and tomato extract analysis. In the general concluding discussion three subjects in particular were clarified:

— the reference materials used should really simulate samples (e.g. for recovery: extraction may be different, some times extractability increases by adding some water to avoid mistakes). Labelled isomers are available for some OPC's, which allow control of analytical steps.

— spiking of samples with solutions (storage? %water? %solvents?) and elimination of solvents also needs great care. Pesticides may be transformed (during extraction as well), and thus the wrong substance may be measured. Intermediate hydrolysation and derivatization is helpful in confirmation. Are there safer derivatization possibilities than using risky diazomethane? Samples from spraying crops or soils (simulating practical use) may be more realistic.

— Which other compounds should be identified (distinction between metabolites and phosphorus acids, etc)?

## 3. BIOLOGICAL EFFECTS OF ORGANOPHOSPHORUS COMPOUNDS, ROLE OF STEREOISOMERS

Prof. Jan H. Koeman (Dept. of Toxicology, Agricultural University, NL-7603 BC Wageningen) presented an excellent introduction to the toxicology of organophosphorus compounds, in relation to occupational, consumer and environmental hazards. To-day about 96 different active components are used as pesticides, plasticizers, flame proof agents, antioxidants, lubricants, etc. One has to distinguish between phosphates, phosphonates, phosphorothionates, phosphorodithionates, TOCPs($= (aryl - O)_3 P = O$), and their toxic derivatives. Already in the 1930s Schrader (Bayer AG) studied systematically the biological effects of OPCs to replace imported pesticides such as nicotine. The main effects are acetyl-cholinesterase inhibition, delayed neurotoxicity, and non-degenerative electromyographical effects (abnormal EMG's, decreased MCV's and increased achilles tendon reflex may be used for controls at working places [early warning!]; for instance those working with OPCs or chlorinated organic compounds have a reduced EMG amplitude). Acetyl-cholinesterase enzyme molecules contain an esteratic and an anionic site, and the OPCs react with serine in these molecules. The reversible product gets irreversible by ageing, thus causing secondary effects (muscarine, nicotinic and CNS-effects, also leading to muscle cramps). While the first effects disappear, second steps are much more critical (distal axonopathy leads to transformations of the Schwann cells and of the Ranvier knots near muscles), and lethal and chronic poisonings occurred in Algeria, the United States of America, and Switzerland. Some OPCs must first be activated to be able to bind to microsomes. Some OPCs are mutagenic in short-term tests, but they are not carcinogenic. Insects react differently from mammals. However, we often have to deal with mixtures of substances of different toxicology. Some substances in technical mixtures may inhibit detoxification (big accidents in Pakistan). Relative rate of dealkylation decreases parallel to $LG_{50}$s in relation to species (rat > mouse > rabbit > dog; rats are especially sensitive). Some OPCs are more toxic to birds than to mammals (if they are transformable into toxic "oxons", which can be degraded by "A"-esterase in mammals, but not in birds).

Dr. Martin K. Johnson (MRC Toxicology Unit, Carshalton, Surrey SM5 4EF, U.K.) reported on the importance of chirality in influencing both acute and delayed neuropathic toxicity of organophosphorus esters. Depending on stability of OPCs, he distinguished between different aspects: identification, intoxication, induced delayed neurotoxicity, and severe defects in walking (ataxia) developing two weeks after dose (through degeneration of distal ends of some axons in spinal cord and in peripheral nerves through NTEs (neuropathy target esterases). With O-P-inhibition esterases cannot reach proteins, but reactivation is possible before ageing. Inhibition of NTEs may lead to initiation of neuropathy, or to protection. Modification of NTEs (after acylation by some OP esters) depends on "ageing", but is not yet clear, what happens (upsetting of other functions? crosslinkings?) with different rates. In analytical chemistry it is important to segregate metabolites and the small fraction reacting with the target (differentiating between metabolic disposal and activation). Before binding to tissues takes place *in vivo*, hydrolysis by plasma must be considered. There are quite a few steps which must be studied:

OPCs→Delivery (eventually detoxification)→primary reaction with target→biochemical response→chemical condition.

In rat liver microsomes 40% of Cyanofenphos stays unchanged, and the ratio (within the other 60%) between exchange of "S" by "O", and hydrolysis may change. The attack on either AChE or NTE by OP esters is also influenced by steric factors. Thus $C(+)P(-)$ isomers are especially toxic in mice (Benshop et al., 1984), while $C(-)P(+)$-Soman inhibits easily NTE ($P(-)$ isomers inhibit rather AChE). Johnson, Read and Benshop (1985) found that the ratio $Ka^{AChE}/Ka^{NTE}$ is a good measure for evaluation OPC-risks (for instance *in vitro* in hen brains). The related reactions of spontaneous reactivation and "ageing" of inhibited esterases are dependent on molecular structure of the phosphyl-enzyme. When NTE is inhibited *in vivo* by the L($-$) isomer of EPN or EPN oxon (4-nitrophenyl ethyl phenylphosphonate), "ageing" of inhibited NTE is rapid and polyneuropathy follows two weeks later; when NTE is inhibited by the D($+$) isomer, ageing is negligible and dosed animal became resistant to the neuropathic effects (*not* the cholinergic effects) of challenge OP esters. Many different effects must thus be considered, and it depends on such detailed information to predict a neuropathic response or protection of OPCs. Chiral analysis in toxicology (during delivery and/or interactions) helps to identify routes and to predict progress and outcome of intoxification (and to explain inter-species differences). Effects of mixtures depend on the dose ratios, and may be different in low and high doses. The critical question is always, which isomer reaches the target first. Leo P. de Jong (Prins Maurits Laboratorium TNO, NL-2280 Rijswijk) talked about an assay for the four stereoisomers of the chiral anticholinesterase Soman (($CH_3)_3CC(H)H_3CO(H_3$)—PO—F) in biological samples. Inhibition rates of the P($-$) isomers are more than 10 000 times greater than those of the P($+$) isomers (at pH 7.5 and 25°C). Separation is possible with capillary GLC (detection is easier than with HPLC) and Carbowax-Chirasil-Val (see also A. Verweij, Section 2) before and after residue inhibition of AChE. It can be seen that the P($-$) isomers disappear more rapidly in the reaction. Treatment with NaF leads to an increase in the enzymatic assay and in the chiral GLC-assay (and reactivation of the P($+$) isomers). Further experiments (also in rat blood on residual Soman concentrations) were also made with deuteriated $D_3$-Soman and $D_{13}$-Soman. With GLC it is possible to separate $H_3C(-)P(+)$, $D_3C(-)P(+)$, $D_{13}C(+)P(+)$, $H_{13}C(+)P(+)$, $D_{13}C(-)P(+)$, and $H_{13}C(-)P(+)$. Studies on reactivation and/or stabilization were also described, when using added amounts. In a general discussion influences of OPCs on biological systems and influences of biological systems on OPCs were distinguished. It must be remembered that only a very small part of OPCs reacts (less than 0.01% of the dose, and that animals can get tolerant by repeated administering). Extrapolations are thus difficult (also in risk assessment). It is unlikely that diet has an influence, but reactivity may also be different in different organs. Another question is, what do enzymatical observations mean to toxicologists.

Mrs Zelimira Vasilic (Institute for Medical Research and Occupational Health, YU-41001 Zagreb) determined metabolites of organophosphorus pesticides in urine as an indicator of occupational exposure. During spraying of an apple-yard urinary concentrations of dimethyl phosphorothiolate potassium salt (for Demeton-S-methyl), and dimethyl phosphorothionate and phosphorodithioate potassium salts (for Azinphos-methyl and Methidathion) were monitored every third day. The author also controlled formulators exposed to a variety of OP pesticides. The author alkylated urinary ether extracts with diazomethane and quantified the derivatives with gas chromatography. She found between less than 20 and about 85 ng/mL of the metabolites, always more of the dialkyl phosphorothionates (which can be determined more accurately) than of the phosphorodithionates. The amounts absorbed were not sufficient to cause a major reduction of enzyme activity (cholinesterase activity in whole blood and plasma was measured for comparison), but the test is suitable for early detection of defectiveness in protection measures.

## 4. GENETICALLY DETERMINED POLYMORPHISM

Dr. T. L. Diepgen (Institute for Medical Statistics and Documentation) and Professor Marika Geldmacher-von Mallinckrodt (German Research Association for Clinical Toxicological Analysis, University of D-8520 Erlangen) evaluated in a lecture and a poster the inter-ethnic differences of the human serum paraoxonase polymorphism. It was found that the serum paraoxonase genotype (hydrolysing paraoxon, and thus influencing paraoxon clearance, and changing with age) shows different patterns in different ethnic groups. With the kinetic Krisch's method, measurements were made of samples around the world (precaution: diethylene glycols may inhibit OPCs). In Caucasians the polymorphism is governed by two alleles. The first allele has a gene frequency $p_{low}$ of 0.67 to 0.78, and is manifested in both the form of a first homozygote group with low activities and a second heterozygotic group with medium activities. About 50% of all Europeans belong to the low activity group, which is still increasing relatively (pregnant women have an even lower activity). The second allele with a gene frequency of $q_{high}$ of 0.22 to 0.33 is manifested in the second heterozygotic and a third homozygotic group with medium resp. high activities. The Hardy–Weinman rule for a two allele model is valid for distribution. The analysis further shows that worldwide, three groups of individuals can be distinguished, and the validity of the Hardy–Weinberg rule must be rejected for Non-Caucasians. The percentage of the low activity group decreases as one moves from Europe to Africa and Asia. In most of the Mongoloids and Negroids only 5–20% of the population can be included in the low activity group, which is not even demonstrable in Aborigines, Maoris, Tonga and some African and Indian (Central America) tribes. In the most rapidly degrading group, the half-time of degradation in serum resp. paraoxon clearing is at least six times shorter than in the other two groups. Paraoxonase total activity is about 50 U/l in Europeans and East Indians, about 200 U/l in American Blacks and Red Indians, and about 300 U/l in African Blacks. Berbers and Ethiopians (mixed populations) show all three groups of activity. Of course resorbitivity is also very different for OPCs. The authors discussed also the G. Carro–Ciampi-method (*Can. J. Pharmacol.* **59**, 904, 1981; *Can. J. Physiol. Pharmacol.* **61**, 336, 1983), and used the ratio (paraoxon hydrolysis rate buffer $+ CaCl_2$/paraoxon hydrolysis rate buffer $+ CaCl_2 + NaCl$) to designate enzyme activation by sodium ions (which is a useful indicator).

ERNEST MERIAN
Therwil, Switzerland.

# PART 1

# TOXICOLOGICAL
# AND
# CHEMICAL ASPECTS

# The Importance of Chirality in Influencing Both Acute and Delayed Neuropathic Toxicity of Organophosphorus Esters†

MARTIN K. JOHNSON

*MRC Toxicology Unit, Woodmansterne Road, Carshalton, Surrey, SM5 4EF, U.K.*

(*Received August 23, 1986; in final form September 21, 1986*)

Triesters (or mixed ester/amides) of phosphoric acid, diesters of organophosphonic acids or monoesters of di-organophosphinic acids may be designed to be stable plasticizers, hydraulic fluids or flame retardants or may be designed as pesticides. In the latter case one ester bond is more labile so that the whole molecule may act as an agent to acylate certain critical targets *in vivo*. Pesticides are often prepared as derivatives of the analogous phosphorothioic acids which are more stable in storage but are metabolized to the active P=O compounds (oxons) *in vivo*.

Typical acute toxicity of OP esters arises from inhibition of vital acetylcholinesterase (AChE) which normally hydrolyses neurotransmitter liberated at neural junctions: acetylcholine (ACh) accumulates and causes excessive stimulation of nerves which can become fatal.

Organophosphorus-induced delayed polyneuropathy (OPIDP) is caused by some but not all OP esters by acylation and subsequent modification of the structure of a neural protein commonly called NTE (neuropathy target esterase). This modification depends absolutely on "ageing" of inhibited NTE.

The attack on either AChE or NTE by OP esters is influenced by steric factors. Thus the inhibitory power of $P(-)$ isomers of soman (pinacolyl methylphosphono-

---

†Presented at the Workshop on Chemistry and Fate of Organophosphorus Compounds, Amsterdam, Holland, June 18–20, 1986.

This article was first published in *Toxicological and Environmental Chemistry*, Volume 14, Number 4 (1987).

fluoridate) against AChE is at least 1000 greater than that of the $P(+)$ isomers. NTE is more easily inhibited by the $C(-)P(+)$ isomer than by the other three. The related reactions of spontaneous reactivation and "ageing" of inhibited esterases are dependent on molecular structure of the phosphyl-enzyme. When NTE is inhibited *in vivo* by the $L(-)$ isomer of EPN or EPN oxon (4-nitrophenyl ethyl phenyl-phosphonate), "ageing" of inhibited NTE is rapid and polyneuropathy follows 2 weeks later; when NTE is inhibited by the $D(+)$ isomer, ageing is negligible and dosed animals become resistant to the neuropathic effects (NOT the cholinergic effects) of challenge OP esters. Other examples are presented.

Analytical methods which distinguish between isomers of free and target-bound agents and their metabolites are necessary to advance detailed mechanistic studies of intoxication by chiral OP esters.

## TOXICITY OF ORGANOPHOSPHORUS ESTERS

There are two main types of organophosphorus (OP) esters which cause toxic effects in man and animals. Firstly, there is the pesticidal type compounds: these are designed to be toxic to at least some living systems and their acute cholinergic effects have been mentioned by Dr. Koeman in his lecture.[1] These esters usually have one hydrolysable bond which is moderately labile either intrinsically or after metabolic activation *in vivo*. Figure 1 shows a number of such compounds. They include the nerve agent soman, parathion and its active metabolite paraoxon, and malathion. The toxicities of these compounds as measured by acute $LD_{50}$ for rats range from about 0.1 mg/kg for sonan, 2–8 mg/kg for parathion and about 1,500–10,000 mg/kg for malathion. However each is obviously a phosphyla-ting agent where phosphylation is a generic term for phosphoryla-tion, phosphonylation, phosphinylation or phosphoramidation. The labile bond has some acid anhydride characteristic and the leaving groups are very diverse.

Pesticidal type OP esters are all developed from the original chemical warfare type such as soman but have features in their molecular architecture which render them susceptible to mammalian metabolic disposal processes. Mammals tend to be more versatile than insects and other pests in their ability to dispose of or destroy OP esters so that malathion, for instance, can be very toxic to insects but much less to animals.

The acute toxicity of OP ester pesticides produces effects ranging from moderate through severe to lethal according to dose. However,

**Figure 1** Common or conventional names and formulae of some OP esters.

| | |
|---|---|
| Soman; pinacolyl methylphosphonyl fluoride | $CH_3[(CH_3)_3.CH.(CH_3).O].P(O).F$ |
| Parathion | $(C_2H_5.O)_2.P.(S).O.(4\text{-}NO_2.C_6H_4)$ |
| Paraoxon; diethyl 4-nitrophenyl phosphate | $(C_2H_5.O)_2.P.(O).O.(4\text{-}NO_2.C_6H_4)$ |
| EPN; ethyl 4-nitrophenyl phenylphosphonothioate | $(C_2H_5.O).C_6H_5.P.(S).O(4\text{-}NO_2.C_6H_4)$ |
| EPN oxon *or* EPNO | $(C_2H_5.O).C_6H_5.P.(O).O(4\text{-}NO_2.C_6H_4)$ |
| 2,2-dichlorovinyl di-*n*-pentylphosphate | $(n\text{-}C_5H_{11}.O)_2.P.(O).O.CH:CCl_2$ |
| 2,2-dichlorovinyl di-*n*-pentylphosphinate | $(nC_5H_{11})_2.P.(O).CH:CCl_2$ |
| Malathion; S-[dicarboxyethyl succinyl] dimethyl phosophorothioate | $(CH_3.O)_2.P.(S).O.CH.C(O).O.C_2H_5$ $\mid$ $CH_2.C(O).O.C_2H_5$ |
| Methamidophos; O,S-dimethyl phosphoramidothioate | $CH_3.O.(CH_3.S.).P(O).NH_2$ |

apart from certain specific effects mentioned later, recovery from non-fatal acute poisoning is usually fairly rapid and apparently complete. There is a suspicion that after convulsive near-lethal doses there may be some subtle long-term neurological effects arising from damage to some brain cells due primarily to localised anoxia during convulsions but this is still debatable.[2]

Compounds of the second type of OP ester which may cause toxicity are apparently much more chemically stable than the insecticidal type. They include various (but not all) symmetrical tri-aryl phosphates and related esters which are favoured as plasticisers, hydraulic fluids and flame retardants. Some of these and some (but, again, not all) insecticidal type cause a delayed onset polyneuropathy called organophosphate-induced delayed polyneuropathy (OPIDP). This syndrome is characterised by the delay of between 1 and 3 weeks after intoxication before clinical effects are manifested. It has often, erroneously, been called OP-demyelinating disease. The better terminology of a "dying-back neuropathy" has in its turn been displaced by the title of distal axonopathy implying symmetrical distal axonal degeneration of certain neurons. The resulting paralysis is not caused by muscle fibre necrosis but by concurrent degenera-tion in distal regions of long, large diameter axons in peripheral nerves and spinal cord.

OPIDP has attracted toxicologists for a number of reasons, of which the commercial importance and extent of potential use of the compounds is but one. Further interest exists because the response can be produced by a single dose in man and various larger mammals and in adult hens, yet there is a delay before frank disorder is seen as if a fuse had been lit but the dynamite keg were a long way removed. Small rodents do not develop clinical neuropathy in response to single doses of typical neuropathic OP esters such as DFP and hens are the laboratory species of choice.

The subject of OPIDP has been reviewed from different angles a number of times starting with Davies in 1963.[3] The reviews in 1964 and 1973 by Cavanagh[4,5] attended principally to pathology. In 1975 I wrote an extensive review of the biochemical mechanism as then understood and explored the areas illuminated by this under-standing.[6] That review was coupled with a further article devoted to structure/activity relationship of every OP compound for which neurotoxicity test data were available.[7] More is now known in molecular terms about the initiation of OPIDP and biochemical studies on the target protein and toxicological applications of the available information have been reviewed recently[8] to show the rationalisation of toxicity testing which has come about as a fruit of understanding. The development of the cellular disorder which follows initiation and precedes clinical expression of neuropathy is still a mystery to be unfolded but understanding the initiation has improved our capacity to suggest rational routes of enquiry.

## ESTERASES AND OP TOXICITY

The four steps of interaction of OP esters with a typical esterase are depicted in Figure 2.

The acute toxicity involves an attack on neural pathways which utilize acetylcholine (ACh) as a neurotransmitter. Signals are usually terminated by destruction of ACh by acetylcholinesterase (AChE) but acutely toxic OP esters inhibit that enzyme by covalent reaction to form inactive phosphylated AChE. Excess ACh then accumulates and toxicity is expressed by prolonged over-stimulation of cholinergic neural systems. In non-fatal cases the intoxication declines at a rate controlled largely by reappearance of a small, but sufficient,

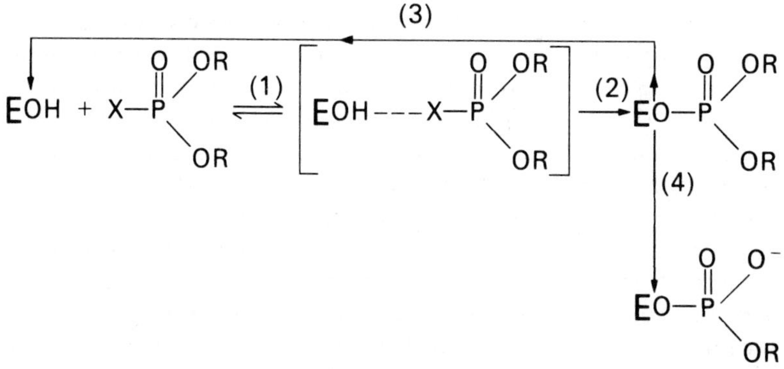

**Figure 2**  Steps in the interaction of an esterase with an organophosphorus inhibitor. (1) Formation of Michaelis complex; (2) Phosphylation of enzyme; (3) Reactivation (spontaneous or forced by oximes or KF); (4) Ageing.

quantity of active AChE either by synthesis or by spontaneous reactivation (dephosphylation; Figure 2) of inhibited enzyme.

For the purposes of this short review it is not necessary to discuss in detail the evidence for the mechanism of initiation of OPIDP. It is sufficient to note that the toxicity involves an esterase commonly called by its acronym NTE which originally stood for "neurotoxic esterase" and now, less confusingly, for neuropathy target esterase. The processes of interaction of OP esters with both AChE and NTE are essentially similar but there are subtle and important differences apart from the fact that each enzyme has its own characteristic tertiary structure which favours interaction with certain subclasses of OP esters. Thus longer-chain dialkyl phosphates have much higher affinity for NTE than do dimethyl phosphates, while for AChE this effect is much less marked.[9, 10]

For AChE the acute toxic effects are identical no matter what the inhibitory molecule which phosphylates (or, indeed, carbamylates or sulphonates) the enzyme in steps (1) and (2) of Figure 2. Duration of intoxication is very dependent on the ease and relative rates of the spontaneous reactivation (3) and ageing (4) reactions. For NTE the picture is different in that two complementary effects can be produced which depend entirely on whether an ageing reaction (step (4) in Figure 2) occurs after inhibition. Thus after inhibition of NTE *in vivo* by 2,2-dichlorovinyl di-*n*-pentyl phosphate (see Figure 1) a typical OPIDP ensues but after inhibition by the analogous phosphinate (Figure 1) not only does neuropathy not develop but dosed animals become resistant to challenge doses of typical neuropathic OP compounds.[11] It is now generally accepted that initiation of OPIDP is a two-step process as depicted in Figure 3 (*X, Y, Z*). Clearly, protective compounds such as NTE-inhibitory phosphinates or phenylmethanesulphonyl fluoride can only proceed through step (1) and the process of initiation is both aborted and blocked. The question has arisen whether ageing follows inevitably after inhibition of NTE by compounds such as phosphates or phosphonates which have an available R—O—P bond which could, on paper, engage in such a reaction. So far as has been tested with alkyl-radiolabelled compounds the ageing reaction of inhibited NTE appears to involve intramolecular transfer of the "R" group to another position on the same polypeptide.[12 – 13a] Clearly, such a reaction may suffer considerable steric constraint. Ageing of inhibited AChE, by contrast,

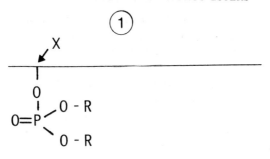

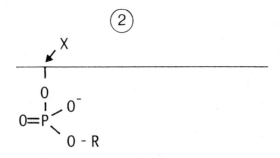

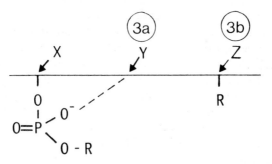

**Figure 3** The proven steps (1 and 2) of initiation of OPIDP with proposed alternative (3a or 3b) consequences. Step (1): Organophosphorylation occurs at $X$ causing inhibition of catalytic activity of NTE; (2) Cleavage of one R—O—P bond to form a negatively charged group with concomitant ageing of inhibited NTE; (3a) Negative charge at $X$ disrupts normal (unknown) function of another site $Y$, OR (3b): transfer of $R$ to a site, $Z$, during ageing disrupts the normal (unknown) function of $Z$. (From Reference 8 with permission.)

appears to proceed by an interaction with solvent and liberation of the "R" group in some form into the medium.

## INFLUENCES OF CHIRALITY ON THE VARIOUS STEPS OF OP INTOXICATION

Prior to interaction with a target, a molecule of toxic compound may have to undergo processes of metabolic activation (usually oxidative, if any, for OP esters) and avoid deactivation and disposal by a variety of routes including other oxidative processes, conjugation or hydrolysis.[2, 14] Since these reactions are enzyme-catalysed, each will have its own structure/activity pattern and there is evidence that enantiomers are handled differently.

### Metabolic disposal of soman

Table I shows data on the preferential destruction of both the P($+$) isomers of soman with comparative sparing of P($-$) isomers on incubation with mouse plasma.[15] This is a hydrolytic reaction but phosphylation of "inessential" esterase sites in liver and elsewhere is also an important route of disposal of soman and other compounds active at around the 1 mg/kg dose level. For this reaction the isomeric preferences are different (Table I) and they are different again for the non-metabolic process developed for isolating P($+$) isomers by Benschop et al.[15] in which the P($-$) are preferentially

**Table I**   Preferential destruction of isomers of soman.

| Procedure | C($+$)P($-$) | C($+$)P($+$) | C($-$)P($-$) | C($-$)P($+$) |
|---|---|---|---|---|
| Hydrolysis in vitro by rabbit plasma | $+$ | $+ + +$ | $+$ | $+ + +$ |
| Binding in vitro to chymotrypsin | $+ + +$ | $+/-$ | $+ + +$ | $+/-$ |
| Binding in vitro to pig liver aliesterase | $+ +$ | $+$ | $+/-$ | $+ + +$ |
| Injected into mice | $+$ | $+ + +$ | $+ +$ | $+ + +$ |

Note: The units of grading are arbitrary and not comparable between procedures. Data from References 15 and 16.

inactivated by reaction with stoichiometric amounts of purified chymotrypsin.

## Metabolic activation of cyanofenphos
## (ethyl 4-cyanophenyl phenylphosphonothioate)

Table II shows the comparative rates of oxidative and hydrolytic reaction of the (+) and (−) isomers of this compound catalysed by rat liver microsomes *in vitro*. Whereas, for one isomer the rates are roughly equivalent, the hydrolytic disposal reaction is dominant over the oxidative activation reaction for the (−) isomer. When it is realised that there are numerous sites in the body where one or other of such reactions may occur (i.e. intestine, plasma, liver, kidney, etc.) and that there is no guarantee that steric preferences will be the same in each tissue, then it is not surprising that extrapolation from observations *in vitro* to predictions of effects *in vivo* is difficult. Moreover it should be noted that the levels and balance of activating and disposing enzymes will not only vary between tissues but between individuals according to their metabolic state and also between species.[2]

**Table II**  Stereoselectivity in metabolism of isomers of cyanofenphos by rat liver microsomes *in vitro*.

| Reaction | Conversion (%) | | |
| --- | --- | --- | --- |
| | (+) Isomer | (−) Isomer | Racemate |
| Oxidation to oxon | 27 | 11 | 20 |
| Hydrolysis to 4-cyanophenol | 26 | 40 | 33 |
| Unchanged | 41 | 42 | 41 |
| Other reactions | 6 | 7 | 6 |
| Oxidation/hydrolysis | 1:1 | 1:4 | |

*Note:* Incubation was for a fixed time in presence of NADPH at pH 7.5 and 37°. Data from Reference 17.

## Inhibition of AChE

Table III shows the rate constants for inhibition of bovine erythrocyte AChE by the four resolved isomers of soman and the appro-

M. K. JOHNSON

**Table III** Anticholinesterase activity *in vitro* and acute toxicity to mice of four stereoisomers of soman.

| Isomer | Inhibition of AChE: $k_a$ $(M^{-1} min^{-1})$ | Approx. $LD_{50}$ (mg/kg) |
|---|---|---|
| C(+)P(−) | $> 10^8$ | 0.1 |
| C(+)P(+) | $< 10^4$ | 7 |
| C(−)P(−) | $> 10^7$ | 0.2 |
| C(−)P(+) | $< 10^4$ | 2 |

*Note:* AChE was from bovine erythrocytes. Data from Reference 15.

ximate $LD_{50}$s for these isomers in mice. The P(−) isomers are 3 to 4 orders more inhibitory than the P(+) although obviously the capacity of all four as simple phosphorylating agents of small molecules is identical. The differences therefore lie in steric considerations of access and "fit" at the catalytic centre of the enzyme. However the huge differences in inhibitory power are only partially reflected in toxicity with the P(−) compounds being a mere 10–70 times more potent. Clearly, metabolic disposal of the isomers must be the dominant influence in determining the toxic dose.

## Comparative inhibitory power against AChE and NTE

Table IV shows the rate constants for inhibition of both enzymes from hen brain by the soman isomers. The differences for NTE are all within a 15-fold range unlike the 1,000-fold for AChE. When the ratio of inhibitory power against the two targets is calculated, it is obvious that, in an intoxication, if a molecule of either P(−) isomer

**Table IV** Comparative reactivities of four stereoisomers of soman against hen brain AChE and NTE *in vitro*.

| Isomer | $k_a^{AChE}$ | $k_a^{NTE}$ | $k_a^{AChE}/k_a^{NTE}$ |
|---|---|---|---|
| C(+)P(−) | $7 \times 10^7$ | $2 \times 10^4$ | 3500 |
| C(+)P(+) | *ca.* $3 \times 10^4$ | $5 \times 10^4$ | 0.6 |
| C(−)P(−) | $4 \times 10^7$ | $5 \times 10^4$ | 800 |
| C(−)P(+) | *ca.* $8 \times 10^4$ | $3 \times 10^5$ | 0.27 |

*Note:* Data from Reference 18.

is delivered intact to the nervous system there will be a marked preference for inhibition of AChE with consequent cholinergic effects while the effect on NTE will be negligible. However if P(+) isomers are delivered, then it seems that there would be a slight preference for attack on NTE. The latter prediction has recently been shown to be true in tests on hens. Interestingly, the attack on NTE has turned out to be a neuropathy protective effect rather than neuropathy-inducing.[19, 19a]

## Spontaneous reactivation and ageing of inhibited AChE and NTE

Table V shows comparative rates of spontaneous reactivation and ageing *in vitro* after inhibition by various chiral compounds. It is difficult to be certain about studies on AChE with P(+) isomers of soman because of the extreme potency of the P(−) isomers which, if present even as only 1% stereo-impurity, would act as the principal inhibitory species. However, the effects of chirality are still easily contrasted between the two enzymes. Thus AChE inhibited by P(−) isomers of soman ages very fast, while ageing of inhibited NTE is so

**Table V**   Spontaneous reactivation and ageing *in vitro* of hen brain AChE and NTE after inhibition by various chiral OP esters.

| Compound | AChE | | NTE | |
|---|---|---|---|---|
| | Reactivation | Ageing | Reactivation | Ageing |
| Isomers of soman | | | | |
| C(+)P(−) | 0 | + + + + | + + | +/− |
| C(+)P(+) | ? | ? | 0 | +/− |
| C(−)P(−) | 0 | + + + + | + + | +/− |
| C(−)P(+) | ? | ? | 0 | + |
| Isomers of analogues of EPNO | | | | |
| D(+) | 0 | 0 | 1 | 0 |
| L(−) | 0 | 0 | 0 | + + + |
| Analogues of methamidophos | | | | |
| D.L. | + + | 0 | 0 | 0 |

*Note:* The units of grading signify conversion of more than 50% within 10 min at 37° (+ + + +) ranging to barely detectable in 18 h (+). Data from References 18, 20 and 21.

small in 18 hr as to be doubtful in significance but, on the other hand, spontaneous reactivation of the inhibited NTE does occur. With P(+) isomers no spontaneous reactivation of inhibited NTE was detected.[18] With resolved S-halobenzylthio esters of isomers of ethyl phenylphosphonic acid analogous to EPN oxon, neither form of inhibited AChE reactivated spontaneously nor did they age.[20] By contrast, one form of inhibited NTE did reactivate rather slowly while the other form aged rapidly.[20] Finally, when AChE was inhibited by a series of unresolved compounds analogous to meth-amidophos, spontaneous reactivation was substantial in 18 hr and ageing was undetected while neither ageing nor reactivation was found with inhibited NTE.[21] This last observation is puzzling in the light of several cases in man of delayed polyneuropathy reported to have been precipitated by intoxication with methamidophos.[22] Possibly, in these cases, the human NTE was inhibited by an isomer other than that which inhibits when the enzyme was inhibited by a racemic compound *in vitro*.

## Delayed neuropathic effects *in vivo*

As mentioned above, one isomer of EPN-oxon produces a form of inhibited NTE which ages but the other isomer produces a non-ageing form *in vitro*. Table VI shows the corresponding effects *in vivo*. The preparations of EPN and EPN oxon were not quite sterically pure but the balance of aged and unaged inhibited NTE

**Table VI** Complementary neuropathic or protective effects of isomers of EPNO and EPN administered to hens.

| Compound | Ratio of isomers in dose L(−)/D(+) | Dose (mg/kg) | Inhibited NTE(%) in aged/unaged form | Neuropathic response | | |
|---|---|---|---|---|---|---|
| | | | | Positive | Negative | Protective |
| EPNO | 90/10 | 14 | 72/4 | 3/3 | 0/3 | — |
| EPN | 90/10 | 58 | 58/6 | 2*/2 | 0/2 | — |
| EPNO | 10/90 | 20 | 10/53 | 0/2 | 2/2 | 4/4 |
| EPN | 10/90 | 75 | 33/55 | — | — | 4/4 |
| EPN | 50/50 | 175 | 46/31 | 9*/66 | 57/66 | — |

*Response in affected birds was moderate rather than severe which seems to require somewhat higher percentage of NTE to be converted to aged inhibited form. Data from References 23 and unpublished.

produced in brain and spinal cord after oral doses of these compounds shows that chirality had been substantially preserved during the processes of absorption, distribution, metabolism and interaction at the target. As a consequence, doses of L(−) EPN oxon or EPN which produced high percentages of aged inhibited NTE also caused OPIDP in pair-dosed birds. Doses of the D(+) isomers produced mostly unaged inhibited enzyme, did not cause neuropathy and were protective when tested with a challenge of neuropathic test compound.[23] When racemic EPN was dosed, then more of the inhibited NTE was in the aged than in unaged form and a small proportion of pair-dosed birds developed neuropathy (Johnson, unpublished).

## CONCLUSION: THE NEED FOR CHIRAL ANALYSIS

Chiral GLC analysis has been mentioned earlier in this paper and the work of Benschop et al.[15] and Nordgren et al.[16] illustrate how one group applied the technique in following previously separated isomers while the other followed two separate components of an isomeric mixture through a biological system. There are several papers in the literature[24–26] reporting separation of chiral isomers of an OP ester by chiral HPLC. The benefit to biological science of further developments in such techniques is apparent. The examples cited in this paper remind us of the crucial importance of steric configuration in the interaction of toxic chemicals and their targets. Studies of biological activity of interesting chiral compounds developed either as potential therapeutic agents or as pesticides or industrial chemicals seldom start with pure stereoisomers except for occasional compounds derived from natural sources. Metabolism and toxicity studies nearly all involve use of the racemic mixture. It is unusual to have large quantities of pure chiral compounds available for testing so that it is important to have analytical procedures which can detect and quantify both chiral products in the presence of each other after administration of racemic compound. It would be a significant step forward if the analytical chemical procedures used to identify circulating material or excreted products could retain and quantify the isomers. This would make it possible to identify the routes of metabolism and also the time-course of activation and disposal of each isomer. Marked differences might occur which could be of major significance for mode of action. Furthermore, studies using both chemical and biochemical techniques would be valuable in establishing the chiral configuration of material bound at target sites. In the case of OPIDP, at least, this configuration appears to determine the outcome of intoxication and it seems unlikely that this is a unique situation.

Information on effects of chirality in molecular processes *in vivo* at all stages after administration of a compound may help to explain inter-species differences and to guide the choice of appropriate species for toxicity tests relevant to the human situation.

## References

1. J. H. Koeman, *Toxicol. Env. Chem.* **00, 000** (1987).
2. International Programme on Chemical Safety. *Environmental Health Criteria for Organophosphorus Pesticides. A General Introduction* (World Health Organization, Geneva, 1986). In press.
3. D. R. Davies, in: *Hanbuch der exp. Pharmakol., Erganzungswerk* (G. D. Koelle, ed.) (Springer, Berlin, 1963), Vol. XV, pp. 860–882.
4. J. B. Cavanagh, *Int. Rev. Exp. Path.* **3**, 219 (1964).
5. J. B. Cavanagh, *Crit. Rev. Toxicol.* **2**, 365 (1973).
6. M. K. Johnson, *Crit. Rev. Toxicol.* **3**, 289 (1975).
7. M. K. Johnson, *Arch. Toxicol.* **34**, 259 (1975).
8. M. K. Johnson, *Rev. Biochem. Toxicol.* **4**, 141 (1982).
9. M. K. Johnson, *Biochem. Pharmacol.* **24**, 797 (1975).
10. M. Lotti and M. K. Johnson, *Arch. Toxicol.* **41**, 215 (1978).
11. M. K. Johnson, *J. Neurochem.* **23**, 785 (1974).
12. B. Clothier and M. K. Johnson, *Biochem. J.* **177**, 549 (1979).
13. B. Clothier and M. K. Johnson, *Biochem. J.* **185**, 739 (1980).
13a. D. G. Williams and M. K. Johnson, *Biochem. J.* **199**, 323 (1981).
14. W. C. Dauterman, *Bull. Wld. Hlth. Org.* **44**, 133 (1971).
15. H. P. Benschop, C. A. G. Konings, J. Van Genderen and L. P. A. De Jong, *Toxicol. Appl. Pharmacol.* **72**, 61 (1984).
16. I. Nordgren, G. Lundgren, G. Puu, B. Karlen and B. Holmstedt, *Fund. Appl. Toxicol.* **5**, S252 (1985).
17. H. Ohkawa, N. Mikami and J. Miyamoto, *Agric. Biol. Chem.* **41**, 369 (1977).
18. M. K. Johnson, D. J. Read and H. P. Benschop, *Biochem. Pharmacol.* **34**, 1945 (1985).
19. M. K. Johnson, J. L. Willems, H. C. De Bisschop, D. J. Read and H. P. Benschop, *Fund. Appl. Toxicol.* **5**, S180 (1985).
19a. M. K. Johnson, J. L. Willems, H. C. De Bisschop, D. J. Read and H. P. Benschop, *Toxicol. Appl. Pharmacol.* In press.

20. M. K. Johnson, D. J. Read and H. Yoshikawa, *Pestic. Biochem. Physiol.* **25**, 133 (1986).
21. M. K. Johnson, E. Vilanova and J. L. Vicedo, *Pestic. Biochem. Physiol.* Submitted for publication.
22. N. Senanayake and M. K. Johnson, *N. Eng. J. Med.* **306**, 155 (1982).
23. M. K. Johnson and D. J. Read, *Toxicol. Appl. Pharmacol.* In press.
24. W. H. Pirkle, J. M. Finn, J. L. Schreiner and B. C. Hamper, *J. Amer. Chem. Soc.* **103**, 3964 (1981).
25. T. M. Brown and J. R. Grothusen, *J. Chromatog.* **294**, 390 (1984).
26. Y. Okamoto, S. Honda, K. Hatada, I. Okomoto, Y. Toga and S. Kabayashi, *Bull. Chem. Soc. Japan* **57**, 1681 (1984).

# Photoreduction of Parathion Ethyl[†]

W. SCHWACK

*Institute for Pharmacy and Food Chemistry, University of Würzburg, Am Hubland, D-8700 Würzburg, F.R.G.*

(*Received June 27, 1986; in final form July 31, 1986*)

Photodegradation of parathion ethyl in presence of unsaturated biomolecules takes place by sufficient reduction of the phenyl nitro group. Main products are N,0-dialkenyl hydroxylamino, cis/trans azo, and azoxy derivatives of parathion.

KEY WORDS: Parathion, photoreduction, azo dimer, azoxy dimer, binding to olefins, transport through membranes, plant cuticles, chromatography.

## INTRODUCTION

Parathion (parathion ethyl, **1**) is one of the most important organophosphorus insecticides and is extensively used in agriculture practice as a contact insecticide. As a very lipophilic compound it is distributed in the outer cuticle layer of plants and hardly penetrates into plants before it is metabolized, e.g. to water-soluble paraoxon.

Therefore, disappearance pathways after application are predominated by plant growth, oxidative and hydrolytic reactions, and by photochemical degradation in sunlight. Photochemical reactions of parathion have been studied several times, preferably in water-containing solutions.[1−4]

†Presented at the IAEAC Workshop on Chemistry and Fate of Organophosphorus Compounds, Free University, Amsterdam, June 18–20, 1986.

This article was first published in *Toxicological and Environmental Chemistry*, Volume 14, Numbers 1 + 2 (1987).

Oxidation to paraoxon, rearrangements to S-ethyl or S-nitro-phenyl phosphates, and hydrolyses affording $p$-nitrophenol and different phosphates readily occur.

On cotton leaves Joiner and Baetcke[5] found paraoxon as the main product, followed by $p$-nitrophenol and O,S-diethyl O-nitrophenyl phosphate.

After spraying, pesticides (especially lipophilic compounds) first of all get in contact with plant cuticles by absorption and distribution therein, directly effected by sunlight.

Under the author's investigations it is of main interest which photoinduced reactions of pesticides occur in presence of bio-molecules of plant cuticles with special view on photoaddition reactions producing "bound residues". In order to establish the photochemical reactivities of 1 in presence of plant cuticle constituents, model photoreactions were undertaken: (1) in presence of cyclohexene as a model for unsaturated compounds of plant cuticles (alkenes, terpenes, sterols) and (2) in presence of methyl oleate as an example of octadecenoic acids which often occur in plant cuticles.


## EXPERIMENTAL

*High performance liquid chromatography:* Degradation rate analyses were carried out on a Pye Unicam hplc system equipped with an automatic sampler (WISP 710P, Waters Associates). Data processing used a Hewlett Packard 3390A integrator. Hyperchrome columns, $250 \times 4.6$ mm, filled with Shandon ODS Hypersil, $5\,\mu$m, (Bischoff, Leonberg) were used. With methanol/water $(75+25)$ parathion was eluted after 6 min.

*Gas chromatography:* A Hewlet Packard GC 5830A gas chromatograph equipped with a flame ionization detector in combination with a 188500A terminal was used. (SE 54 fused silica capillary column).

*Mass spectra* (EI mode) were recorded on a CH7/Varian MAT system.

*$^1$H-nmr spectra* were recorded on a Bruker WM 400 spectrometer in CDCl$_3$ and reported in ppm downfield from TMS as internal standard.

*IR spectroscopy* was carried out with a Beckman IR 4240 spectrometer.

*Photolyses*: 100 mg parathion (0.34 mmol) dissolved in 10 ml cyclohexene (p.a., dest. over $P_2O_5$, Merck, Darmstadt) or methyl oleate (99%, Sigma Chemical Co.) was irradiated in a quarz tube for 10 h using a 150 watt high pressure mercury lamp (TQ 150, Hanau Quarzlampen GmbH) equipped with a quarz glass water-cooling jacket. The uv light was filtered by a glass filter WG 295, cut-off $\lambda < 280$ nm, (Schott Glaswerke, Mainz) before reaching the samples.

*Product isolation*: The cyclohexene photolysis mixture was evaporated to dryness, dissolved in 2 ml diethyl ether, and chromatographed on silica gel (LOBAR B column, LiChroprep, Si 60, 0.06–0.04 mm, Merck, Darmstadt) using petroleum ether/diethyl ether (90/10) as eluant. For light protection the column was covered with aluminia foil. Fractions were detected by a uv detector (254 nm). The photolysis mixture in methyl oleate was prechromatographed on 100 g silica gel (0.063–0.2 mm, Woelm) using petroleum ether containing 2% diethyl ether to separate methyl oleate. Elution with petroleum ether/diethyl ether (80/20) recovered the product mixture which was chromatographed as above.

*0,0-diethyl    0-(4-(N,0-bis(cyclohex-2-ene-1-yl)-hydroxylaminophenyl))* *phosphorothioate* (**2**):    Colourless oil.

MS (70 eV) m/z = 341 (69%, $M^+ - C_6H_8O$); 313 (20%); 261 (36%); 233 (12%); 205 (20%); 125 (42%); 109 (25%); 108 (27%); 97 (33%); 81 (100%); 79 (33%); 65 (25%); 53 (31%); 41 (31%).

IR (neat)    3025, 2980, 2940, 2870, 2840, 1645, 1490, 1390, 1200, 1160, 1090, 1025, 940, 840, 820, 720 cm$^{-1}$.

$^1$H-NMR (400 MHz, CDCl$_3$)    $\delta = 7.07$–7.15 (4 H, AA′BB′, aromatic protons); 5.67–5.89 (4 H, 2 m, 2—CH=CH—); 4.19–4.27 (4 H, 2 q, 2—OCH$_2$CH$_3$, $J = 7$ Hz); 4.08–4.17 (1 H, m, —O—CH<); 3.83–3.90 (1 H, m, —N—CH<); 1.6–2.1 (10 H, m); 1.46–1.55 (2 H, m); 1.32–1.38 (6 H, 2 t, 2—OCH$_2$CH$_3$, $J = 7$ Hz). C$_{22}$H$_{32}$NO$_4$PS (437, 55).

*trans 4.4′-bis((diethoxyphosphinothioyl)oxy)azobenzene* (**3**):    Orange-red needles, m.p. 61°C (*n*-hexane).

MS (70 eV)   m/z = 518 (38%, M⁺); 273 (6%); 245 (100%); 217 (7%); 189 (13%); 171 (4%); 153 (4%); 125 (42%); 109 (12%); 97 (38%); 93 (38%); 65 (10%).

IR (KBr)   3000, 2950, 2920, 1600, 1590, 1490, 1390, 1235, 1210, 1170, 1150, 1100, 1030, 930, 855, 825, 800 cm⁻¹.

¹H-NMR (400 MHz, CDCl₃)   $\delta = 7.89$–7.92 and 7.31–7.35 (4 H each, AA'BB', aromatic protons); 4.23–4.31 (8 H, 2 q, 4 —OCH₂CH₃, J = 7 Hz); 1.37–1.41 (12 H, 2 t, 4 —OCH₂CH₃, J = 7 Hz).

C₂₀H₂₈N₂O₆P₂S₂ (518, 54) calc.: C (46.33%), H (5.44%), N (5.40%); found: C (46.48%), H (5.46%), N (5.31%).

$\lambda_{max}$ (iso-octane)   232, 330, 440 nm.

$\lambda_{max}$ (95% ethanol) (log ε)   230 (4.22), 330 (4.46), 430 (3.00), 453 (3.00).

*cis 4,4'-bis((diethoxyphosphinothioyl)oxy)azobenzene* (**4**):   Red oil, mass spectrum identical with that of **3**.

$\lambda_{max}$ (iso-octane)   246, 297, 440 nm.

*Photoinduced (daylight) equilibrated isomers mixture of 3 and 4.*

$\lambda_{max}$ (95% ethanol) (log ε)   228 (4.18), 326 (4.31), 430 (3.08), 453 (3.02).

tlc (SiO₂)   petroleum ether/ether (80/20).

tlc (RP18)   methanol/water (90/10).

hplc (RP18)   methanol/water (80/20), detector: 330 nm.

*4,4'-bis((diethoxyphosphinothioyl) oxy)azoxybenzene* (**5**):   Orange-red crystals, m.p. 64°C (ether).

MS (70 eV)   m/z = 534 (60%, M⁺); 518 (16%, M⁺ − O); 289 (5%); 261 (14%); 245 (88%); 233 (8%); 217 (12%); 205 (45%); 189 (20%); 177 (10%); 153 (24%); 109 (44%); 97 (100%); 93 (26%); 65 (22%).

IR (KBr)   3000, 2950, 2920, 1590, 1570 (—NO = N—), 1490, 1390, 1220, 1200, 1150, 1020, 925, 845, 820, 800 cm⁻¹.

¹H-NMR (400 MHz, CDCl₃)   $\delta = 8.30$ and 8.23 (2 H each, d,

aromatic protons, $J = 9\,Hz$); 7.28–7.33 (4 H, 2 d, aromatic protons, $J = 9\,Hz$); 4.22–4.31 (8 H, 4 q, 4—$OCH_2CH_3$, $J = 7\,Hz$); 1.36–1.41 (12 H, 4 t, 4—$OCH_2CH_3$, $J = 7\,Hz$).

$C_{20}H_{28}N_2O_7P_2S_2$ (534, 54) calc.: C (44.94%), H (5.28%), N (5.24%); found: C (45.03%), H (5.34%), N (5.14%).

*Synthesis of* 3: (All works have to be done with good light protection) 1.07 g (5 mmol) 4,4′-dihydroxy azobenzene was introduced into a solution of 1.23 g (11 mmol) potassium tert.-butanolate in 20 ml tert.-butanol. 2.07 g (11 mmol) 0,0-diethyl phosphorochlorothioate (Aldrich), dissolved in 10 ml tert-butanol, was dropped into the solution over a period of 15 min under stirring and cooling with water (20°C). After further 30 min of stirring, the reaction mixture was poured into ice/water and extracted with 100 ml ether. The organic phase was washed with water and 0.1 n NaOH until the washing solutions were colourless. After drying with sodium sulfate, the ether phase was evaporated supplying a dark red oil. Crystallization from *n*-hexane yielded 2.2 g (84%) of orange-red needles, m.p. 61°C.

## RESULTS AND DISCUSSION

As the experiments of Meallier *et al.*[2] had shown, photodegradation of parathion in cyclohexane was very insufficient (Figure 1). But in presence of the olefinic model cyclohexene 1 was rapidly photodegraded as shown in Figure 1.

As a result of intensive colouring of the irradiated solution over yellow to dark-red the violent reaction is visible, too.

Only one main product was detected by gc analysis. After column chromatography a colourless oil (2, 16%) was obtained which ir spectrum showed no absorption bands for the parathion nitro group, indicating that the reaction had taken place there.

Beside the signals for two ethoxy groups and four aromatic protons, there are signals for two cyclohexene rings in the [1]H-nmr spectrum of 2, which means that the parathion skeleton didn't change but that the nitro group is reduced by addition of two cyclohexene molecules.

During these investigations, Draper and Casida reported that

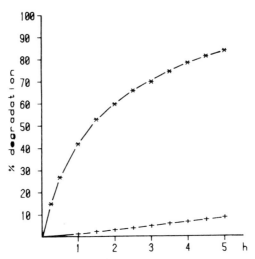

FIGURE 1 Photodegradation rate of parathion in cyclohexane ($-+-+-$) and cyclohexene ($-*-*-$), 50 mg/50 ml ($\lambda > 280$ nm).

nitroaryl pesticides (as well as parathion) can be photoreduced to nitroso derivatives which undergo "ene"-type additions to olefins, yielding alkenylaryl hydroxylamines. These autoxidize readily to stable alkenylaryl nitroxide radicals (Figure 2) as evident by ESR spectroscopy.[6]

The isolated photoproduct 2 fits well in the mechanism proposed by Draper and Casida. Parathion is photoreduced to the nitroso stage 1a which adds to cyclohexene via "ene"-type addition (2a). The resulting nitroxide radical 2b is further added to cyclohexene, yielding the hydroxylamine ether 2 (Figure 2).

In addition to the results of Draper and Casida, this reaction pathway provides further possibilities for nitroaryl pesticides to produce "bound residues" in plants, not evident by ESR.

After irradiation of 1 in cyclohexene, three more reaction products could be isolated as red coloured oils, two of which crystallize after cooling, and which were all not enough volatile for detection by gc. From the view of their spectra these are parathion reduction products, two cis/trans isomer azo dyes 3/4 (36%/12%) and the azoxy product 5 (10%) (Figure 3).

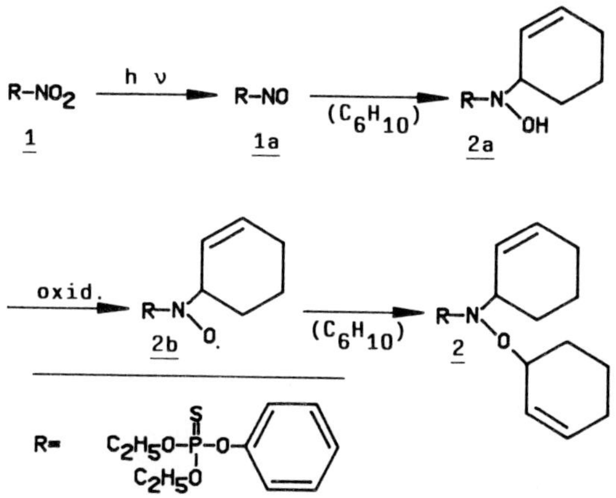

FIGURE 2   Parathion photoreduction to the nitroso derivative and covalent binding to the C—C-double bond of cyclohexene.

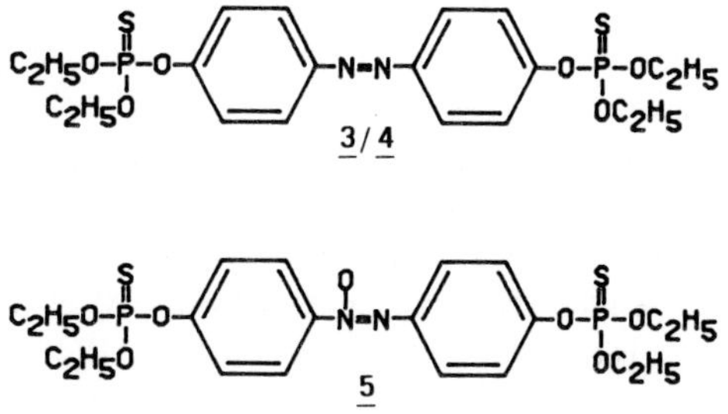

FIGURE 3   Parathion photoreduction products: cis/trans azo dimer (3/4) and azoxy dimer (5).

In solution, both isolated azo dyes photoinduced isomerize (cis↔ trans) under the influence of daylight or artificial laboratory light until equilibrium, so that their isolation and crystallization must be done carefully under light protection. Otherwise the azo products isomerize during column chromatography and are continuously eluted within all fractions between them.

Joiner and Baetcke[5] described their problems during column chromatography where in some instances radioactivity was continuously eluted from the origin to near the solvent front.

Koivistoinen and Meriläinen[7] reported a similar phenomenon with the explanation that there are many rearrangement products present which formed a continuous streak of $^{14}$C-radioactivity. Possibly the formation of photoinduced isomerizing azo products is an additional explanation for the reported problems during isolation of (radio-labeled) parathion photoalteration products.

On uv irradiation of **1** in presence of methyl oleate the reaction mixture even changed to a dark-red colour. The azo and azoxy products **3**, **4**, and **5** could be isolated in the same way (even in some higher yields) as in the case of cyclohexene (45%, 15%, and 18% respectively).

In the course of the work presented in this paper Haffner et al.[8] published their results about the influences of different insecticides on apple quality. They established that there were more carotenoids in apple peels after use of parathion methyl as insecticide, analysed by photometry of the extracted caratenoid fraction, than by plant protection with other insecticides. With the own results in mind it was proposed that there have not been more carotenoids in the apple peels, but orange-red-coloured azo dyes derived from parathion methyl. In order to check this hypothesis, apples were sprayed with pure parathion (dissolved in diethyl ether) as well as with "E 605 forte" (Bayer, dissolved in water), irradiated with filtered uv light and analyzed in the same manner used by Haffner et al. (Kaether).[9] In both cases, the extinctions of the carotenoids extracts of the parathion treated apples were remarkably higher than those of the untreated apples (Table I).

TABLE I

| Treatment | Extinction$_{corr.(453 nm)}$[a] |
|---|---|
| no | 0.13 |
| pure parathion[b] | 0.28 |
| E 605$^R$ forte[b] | 0.22 |
| azo dimer **3**[c] | 0.16 |

[a]Extinction (453 nm) of carotenoids extracts of apples (golden delicious), determined and corrected with the method of Kaether[9] (related to 100 cm$^2$ peels).
[b]1 mg parathion/apple, 5 h uv irradiation ($\lambda > 280$ nm).
[c]0.4 mg/apple added.

After soxhlet extraction of parathion treated apple peels with *n*-hexane, the azo products could be identified by tlc and hplc. It is planned to investigate these establishments in a field experiment next.

For toxicological and insecticidal studies the azo product **3** has been synthesized by reaction of 4,4′-dihydroxy azobenzene with 0,0-diethyl phosphorochlorothioate affording the trans product as orange-red needles in good yields. There is a US patent (1977) for a similar product starting from 4,4′-dihydroxy azobenzene and 0,0-dimethyl phosphorochlorothioate, yielding an azo product derived from parathion methyl. It was effective in controlling fleas.[10] The azo products **3/4**, as thin film on glass plates, are toxic to flies. Further questions for the insecticidal, microbiocidal, and especially for the toxicological characteristics are still under investigation in a separate work, parallel with photochemical syntheses of azo dimers derived from other nitroaryl pesticides.

## CONCLUSIONS

From the isolated products a total photoreduction sequence of parathion in presence of unsaturated biomolecules of plant cuticles is derivable:

$$-NO_2 \rightarrow -NO \rightarrow -NHOH \rightarrow NH_2$$
$$\mathbf{1} \qquad \mathbf{1a} \qquad \mathbf{1b} \qquad \mathbf{1c}$$

$$\mathbf{1a} + \mathbf{1c} \rightarrow \mathbf{3/4}; \quad \mathbf{1a} + \mathbf{1b} \rightarrow \mathbf{5}; \quad \mathbf{1a} + \text{cyclohexene} \rightarrow \mathbf{2}.$$

The first reduction stage, the nitroso derivative **1a**, is the central product. Condensation with the amine **1c** leads to the azo compounds **3** and **4**, while reaction with the hydroxylamine **1b** provides the azoxy dimer **5**. The reaction with olefins is illustrated in Figure 2.

Nevertheless, **1a** and **1c** couldn't be isolated from the reaction mixtures, possibly due to their fast interactions one with another or with olefins. The hydroxylamine **1b** could be proved by tlc and spraying with Tollen's reagent. After irradiation of **1**, in either cyclohexene or methyl oleate some more minor products are found that are still under investigation in order to get a general view of the parathion photoreactivities in presence of unsaturated biomolecules.

## Acknowledgements

Many thanks are expressed to Bayer AG, department plant protection (Monheim), for the gift of *pure parathion ethyl* (min. 99.2%).

## References

1. J. R. Grunwell, *J. Agric. Food Chem.* **21,** 929 (1973).
2. P. Meallier, J. Nury, B. Pouyet, C. Coste and J. Bastide, *Chemosphere* **12,** 815 (1983).
3. J. E. Woodraw, D. G. Crosby and J. N. Seiber, *Res. Rev.* **85,** 111 (1983).
4. M. Mansour and F. Korte, *Bull. Environ. Contam. Toxicol.* **30,** 358 (1983).
5. R. L. Joiner and K. P. Baetcke, *J. Agric. Food Chem.* **21,** 391 (1973).
6. W. M. Draper and J. E. Casida, *J. Agric. Food Chem.* **33,** 103 (1985).
7. P. Koivistoinen and M. Meriläinen, *Acta Agr. Scand.* **12,** 267 (1963).
8. K. Haffner, G. Bünemann and D. Schenker, *Gartenbauwiss.* **50,** 177 (1985).
9. K. E. Kaether, *Gartenbauwiss.* **30,** 361 (1965).
10. M. St. Schrider and St. D. Levy, *U.S. 4,052,381* (C.A. 87: 201087d (1977)).

# The Polymorphism of the Human Serum Paraoxonase[†]

## M. GELDMACHER-VON MALLINCKRODT and T. L. DIEPGEN

*Institut für Rechtsmedizin and Institut für Medizinische Statistik und Dokumentation der Universität Erlangen-Nürnberg, FRG*

(*Received August 4, 1986*)

The activity of paraoxonase, the enzyme which hydrolyses paraoxon, 0,0-diethyl-0-4-nitrophenylphosphate, in human serum shows a genetically determined polymorphism with strong interethnic differences. The serum paraoxonase genotype has a significant influence on the paraoxon clearance. Individuals with high serum paraoxonase activity may be better protected against the toxic effects of parathion (0,0-diethyl-0-4-nitrophenylthiophosphate).

In Caucasians the polymorphism is governed by two alleles. The first allele has a gene frequency $p_{low}$ of 0.67 to 0.78, and is manifested in both the form of a first homozygotic group with low activities and a second heterozygotic group with medium activities. About 50% of all Europeans belong to the low activity group. The second allele with a gene frequency $q_{high}$ of 0.22 to 0.33 is manifested in the second heterozygotic and a third homozygotic group with medium resp. high activities. The Hardy–Weinberg rule for a two allele model is valid for the distribution.

The percentage of the low activity group decreases as one moves from Europe to Africa and Asia. In most of the Mongoloids and Negroids only 5 to 20% of the population can be included in the low activity group, which is not even demonstrable in Aborigines, Maoris, Tonga and some African and Indian (Central America) tribes. The validity of the Hardy–Weinberg rule for a two- or three-allele model must be rejected in non-Caucasians.

---

†Presented at the Workshop on Chemistry and Fate of Organophosphorus Compounds, Amsterdam, Holland, June 18–20, 1986.

This article was first published in *Toxicological and Environmental Chemistry*, Volume 14, Number 3 (1987).

## INTRODUCTION

Organophosphates are esters or amides of phosphoric, phosphonic and phosphinic acid.[1] They are strong inhibitors of acetyl-cholinesterase. The symptoms of poisoning are explained by the accumulation of released acetylcholin at the receptor site. The toxicity of organophosphates is very variable. This is partly caused by the fact that in warm blooded animals some compounds are more rapidly detoxicated than others. Also a toxification through metabolic conversion is observed. The velocity of the detoxification considerably depends on the activity of ester cleaving enzymes.

One of the oldest and most well-known organophosphates is parathion (E 605). In warm blooded animals the relatively untoxic parathion is converted to the PO-analogon paraoxon which is responsible for the toxicity. Parathion and paraoxon are detoxified by hydrolysis[2] (Figure 1).

Most of the detoxification reactions of foreign compounds take place in the liver. On the other side for some poisons also enzymes in the serum play an important role. This is especially true for the hydrolytic cleavage of paraoxon into diethylphosphoric acid and 4-

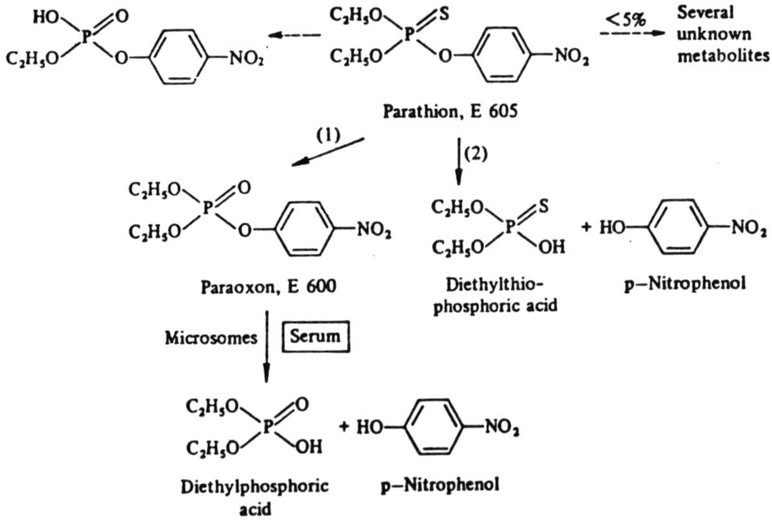

Figure 1   Metabolism of parathion in warm-blooded animals.[2]

nitrophenol by the enzyme paraoxonase, first observed by Aldrige[3,4] in sera of warm blooded animals.

For humans this was shown by Erdös et al.[5-7] as well as by Skrinjaric-Spoljar and Reiner.[8] However, these authors suggested that not only one single enzyme is responsible for the paraoxon hydrolysis in serum. Most of the activity of human serum para-oxonase is found in Cohn fraction IV-1. This enzyme activity depends on the presence of calcium ions and is consequently inhibited by EDTA. A second paraoxonase independent of calcium and consequently not inhibited by EDTA is found in the albumin fraction (Cohn fraction V). According to the enzyme nomenclature the paraoxonase is an arylesterase (E.C. 3.1.1.2).

For the EDTA sensitive part of the human serum paraoxonase activity a polymorphism could be established in West Germans.[9-13] The results of Zech and Zürcher are shown in Figure 2.

During extended family investigations in West Germans we were able to show that this polymorphism is caused by genetic factors and follows a two allele model.[11,12,14]

Playfer et al.[15] confirmed this for a British (Figure 3), Eiberg and Mohr[16] for a Danish, and Carro-Ciampi et al.[17,18] for a Canadian sample.

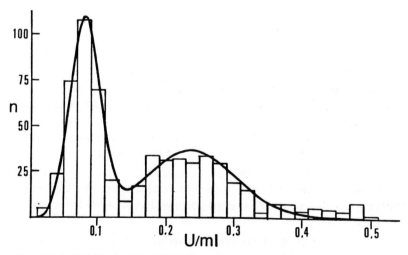

**Figure 2** Individual phosphorylphosphatase (paraoxonase) activity in serum of 619 humans, males and females, the age varying from 4 to 78 years.[13]

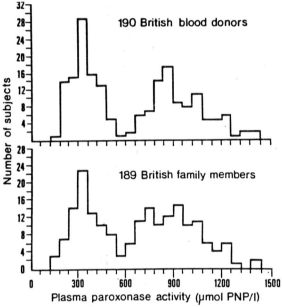

**Figure 3** Frequency distribution histograms of British blood donors and British families.[15]

For the monitoring of workers exposed to parathion and related organophosphates[19] the determination of paraoxonase activity has the following meaning:

1) Paraoxonase definitely protects cholinesterase. This was shown by Geldmacher-von Mallinckrodt et al.[11] as well as Eckerson et al.,[21] who used a mathematical model. From that it can be expected that individuals with high paraoxonase activity at a similar exposition characteristic have a less decreased serum cholinesterase activity than individuals with lower serum paraoxonase activity.

2) Subjects with high paraoxonase activity may at a similar exposition eliminate more paraoxon metabolites (p-nitrophenol, diethylphosphoric acid) in urine than subjects with lower para-oxonase activity.

3) If a monitoring of exposed individuals not only for parathion in blood but also for paraoxon levels is required the blood has to be put immediately in tubes containing EDTA (also citric acid or

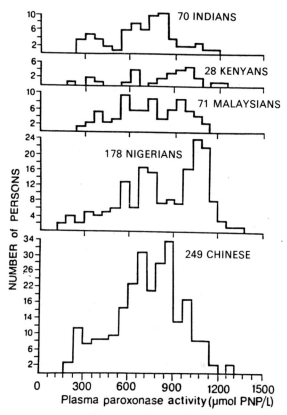

**Figure 4** Frequency distribution histograms of different ethnic groups.[15]

oxalate can be used). If this is not performed, paraoxon will be cleaved in a very short time.

Playfer *et al.*[15] found different patterns of distribution of the serum paraoxonase activity in different ethnic groups (Indians, Kenyans, Malaysians, Nigerians, Chinese: Figure 4). We followed up the problem and investigated additional samples from all over the world. The goal of our studies was the further exploration of the genetic polymorphism of human serum paraoxonase in the light of interethnic differences.

## SOURCES OF THE EXAMINED SAMPLES

All the test subjects were healthy (with the exception of minor dermatologic and gynecologic disorders in the Ethiopians), and did not to their knowledge suffer from any metabolic diseases or liver disorders.

The origin of the serum samples is listed below in alphabetical order according to population:

| | |
|---|---|
| Aborigines | Pure-blooded Aborigines of several reservations near Alice Springs in Central Australia. |
| Afghans | From different tribes in Afghanistan. |
| (U.S.) American Negroes | American soldiers stationed in the Federal Republic of Germany. |
| Berbers | From Tunis and surroundings. |
| Eskimos | Outpatients of the hospital of Angmagssalik, Greenland. |
| Ethiopians | Outpatients with various dermatologic diseases (not including leprosy) of the All-Africa Leprosy and Rehabilitation Centre, as well as gynecologic patients of the St. Paul's Hospital in Addis Ababa. |
| Filipinos | Igorots of the Kordilleren of North-Luzern. |
| Finns | From Oulo. |
| Germans | From various parts of the Federal Republic of Germany. |
| Ghanaians | From Accra and surroundings. |
| Greeks | From various parts of Greece. |
| Hungarians | From various parts of Hungary. |
| Indians | From Bombay and surroundings. |
| Indonesians | Indonesian students in Berlin, almost all of whom were ethnic Chinese. |
| Iranians | From Teheran and Jazd. |
| Italians | From Siena and surroundings. |
| Jamaicans | Outpatients of the Cornwall Regional Hospital in Montego Bay. |
| Japanese | From Osaka and surroundings. |
| Malaysians | Bataks from the lake Toba, island Samosir. |
| Maltese | Blood donors from Malta. |

| | |
|---|---|
| Maoris | Maoris from Auckland, New Zealand. |
| Mexican Indians I | From San Cristobal de las Casas, Estado de Chiapas. |
| Mexican Indians II | From Cuetzalan. |
| Nigerians | Ibos from Iboland. |
| Palestinians | From Nablus, Bier Ziet, Jenin and Rama-Allah (West Bank of Jordan). |
| Rhaeto-Romanics | Sutselvic and Surmeiric Rhaeto-Romanics from Switzerland. |
| Rumanians | From Bucharest and surroundings. |
| Sardinians | Blood donors of Ospedale Civile in Sassari. |
| Senegalese | Sérér from Tattaguine near Fatick. |
| Sri-Lankans | From Colombo. |
| Swedes | Blood donors of Södersjukhuset and Karolinskasjukhuset in Stockholm. |
| Tongans | From Nuku'alofa and surroundings of the island Tongatapu. |
| Vietnamese | Mostly ethnic Chinese who immigrated to the Federal Republic of Germany in 1980. |
| Yugoslavians | Students from Ljubljana. |
| Zambians | Tonga negroes from Mazabuka. |
| Zimbabweans | Outpatients of the M'pilo Hospital, Bulawayo, Matabeleland. |
| Zulus | Blood donors of the St. Mary's Hospital, Zululand. |

## DETERMINATION OF THE TOTAL AND THE EDTA-STABLE ACTIVITY OF THE SERUM PARAOXONASE

The measurements were conducted using the kinetic test of Krisch,[9] in which the enzymatic hydrolysation of $p$-nitrophenol from paraoxon in glycine buffer at a pH of 11.2 and 25°C was photometrically determined.

Contrary to the method suggested by Krisch,[9] our samples contained no ethylene glycol because we found that this substance inhibits paraoxonase. For the measurement of the activity of the EDTA-stable enzyme the glycine buffer contained $10^{-3}$ M EDTA.

## STATISTICAL METHODS

Age, sex, population and individual index and the activity of human serum paraoxonase per test subject were documented and stored for statistical analysis.

Hommel[22] described the mathematical model for separation of a population into groups according to enzyme activities assuming a two-allele model. On the basis of this model the computer program GENITER[23] was developed which is suitable for the iteration of a two- or three-allele model.

This program is available from the Institute for Medical Statistics and Documentation of the University of Erlangen-Nürnberg (Director: Prof. Dr. L. Horbach) for similar studies.

To analyze possible sex related differences in activity the Wilcoxon-test was used.

## TOTAL SERUM PARAOXONASE ACTIVITY IN EUROPEAN POPULATIONS

The histograms of the individual populations (Figure 5) are very similar and show a first peak of low activity.

In all of the European samples, the first group with low activity is the largest, with a relative frequency of about 50%.

It was evident that the hypothesis predicting the validity of the Hardy–Weinberg rule for a two-allele model governing the inheritance of human serum paraoxonase in all European samples can be accepted on the basis of the observed values ($\alpha \leqq 1\%$).

The gene frequencies of the total activity in European samples is shown in Table I.

We also investigated families, i.e. the activities of paraoxonase of father, mother and at least one child from Germany, Finland, Crete, Spain. The results computed by iteration technique with and without children do not differ and are very similar to those of the non-related persons from the same country.

## TOTAL SERUM PARAOXONASE ACTIVITY IN NON-EUROPEAN POPULATIONS

Turks, Palestinians, Afghans, Iranians, Indians and Sri-Lankans have

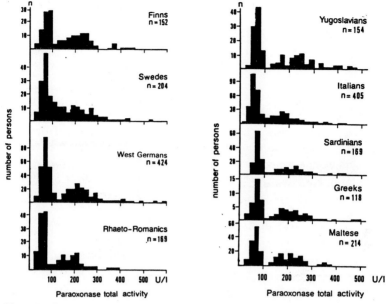

**Figure 5**  Total serum paraoxonase activity in different European samples (examples).

**Table I**  Gene frequencies $p_{low}$ and $q_{high}$ of total serum paraoxonase activity in Europeans.

| Europe: Non-related persons | $p$ | $q$ |
|---|---|---|
| *Central and North Europe* | | |
| 152 Finns | 0.73 | 0.27 |
| 204 Swedes | 0.67 | 0.33 |
| 424 Germans | 0.72 | 0.28 |
| 169 Rhaeto-Romanics | 0.78 | 0.22 |
| *South Europe* | | |
| 46 Hungarians | 0.76 | 0.24 |
| 222 Rumanians | 0.69 | 0.31 |
| 154 Yugoslavians | 0.74 | 0.26 |
| 405 Italians | 0.70 | 0.30 |
| 169 Sardinians | 0.75 | 0.25 |
| 118 Greeks | 0.73 | 0.27 |
| 214 Maltese | 0.76 | 0.24 |
| *Europe: Families* | | |
| 49 Finnish families | 0.78 | 0.22 |
| 65 German families | 0.78 | 0.22 |
| 22 Greek families | 0.78 | 0.22 |
| 28 Spanish families | 0.72 | 0.28 |

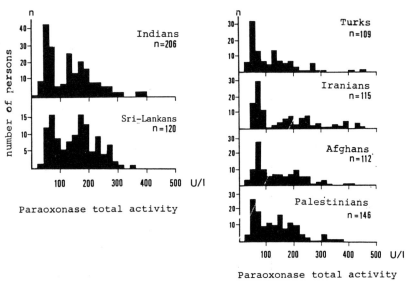

**Figure 6**  Total serum paraoxonase activity in different non-European samples, where we found a large group of low enzyme activity.

a large group of low activity, which is very similar to that found in Europeans (Figure 6).

In Mongoloids and Negroids (Figure 7) there was a very similar distribution which compares to Playfer's data for Chinese.[15] As did the European samples, the Mongoloid and most of the Negroid samples also showed an easily distinguishable group with low activity. The percentage of the low activity groups was significantly lower than in the Europeans. This became especially apparent when the data were analysed statistically using the method of iteration. Convergence of frequency, mean and variance of the group of low activity could be computed very fast—already after some steps—but the results of the other two groups remained constant only after a lot of iteration steps. As well the hypothesis predicting the validity of the Hardy–Weinberg rule for a two-allele model could not be accepted on the basis of the observed values in Mongoloids and Negroids.

In some of the studied samples we could not detect the low activity group neither by eye nor by the method of iteration (Figure 8).

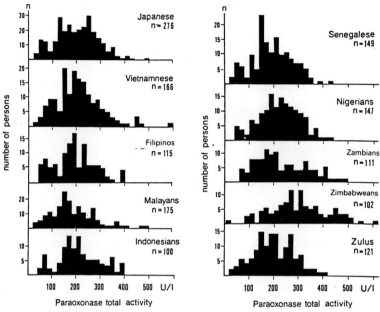

**Figure 7** Total serum paraoxonase activity in Mongoloid and Negroid samples (examples).

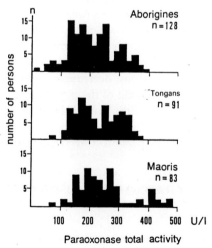

**Figure 8** Total serum paraoxonase activity in samples from the South Pacific area.

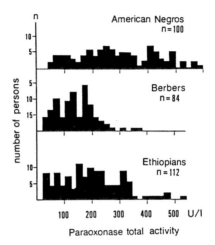

**Figure 9**   Total serum paraoxonase activity in mixed populations.

Also in a few African tribes as well as in an Indian tribe from Central America the low activity group was completely absent.

Also of interest are the investigations in mixed populations like Berbers and Ethiopians as well as in Afro-Americans (Figure 9).

When the frequency of the low activity group is shown in a map of the world the picture shown in Figure 10 is seen. The highest percentage of the low activity group can be observed in Europe. Beginning in Europe this group decreases steadily which is possibly caused by migration and mixing of the populations.

We also investigated families from India, Vietnam, Senegal and Nigeria. The results computed by iteration technique are very similar to those of the non-related persons from the same ethnic group.

## EDTA STABLE PARAOXONASE IN DIFFERENT ETHNIC SAMPLES

In all the samples mentioned so far we have also studied the distribution of the EDTA stable paraoxonase activity. The activity was much lower between 0 and 80 U/l measured according to

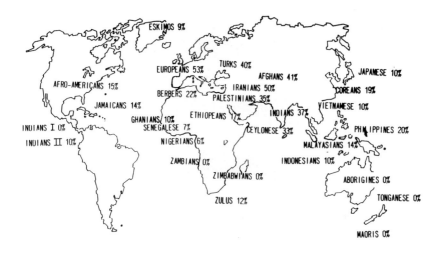

**Figure 10**   Percentages of phenotypes with low total serum paraoxonase activity in random samples taken from different ethnic groups.

Krisch. In all samples a normal distribution became apparent (Figure 11).

## PARAOXON CLEARANCE

To obtain an idea of the extent of the paraoxon clearance in serum we determined the paraoxonase activity in 30 persons using the method of Krisch. The sera of the same probands were incubated with paraoxon (pH 7.4) and the decay of paraoxon was determined by gas-chromatography. An exponential function has been obtained (Figure 12). We found a linear relationship between paraoxon deterioration and paraoxonase activity ($r = 0.95$).

The half-lives under these conditions were between 60.4 and 8.6 min exhibiting significant differences. Sera with low paraoxonase activity have relatively high half-lives and sera with high activity have a very short half-life of 8–10 min. This clearly shows that human paraoxonase under physiological conditions can hydrolyse

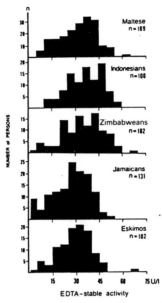

**Figure 11**  EDTA-stable paraoxonase activity in different ethnic groups (samples).

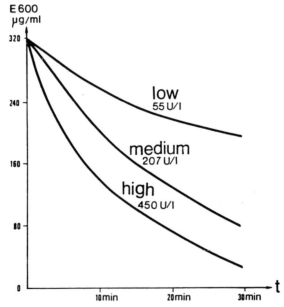

**Figure 12** Degradation of paraoxon in serum. Low = serum with low activity (55 U/l); medium = serum with medium activity (207 U/l); high = serum with high activity (450 U/l).

considerably amounts of paraoxon in individuals with high paraoxonase activity.

## CONCLUSIONS

Our experimental research on human serum paraoxonase has revealed the existence of several distinct enzymes in the serum of each of the test subjects studied which are of quantitative importance for the degradation of paraoxon, the toxic metabolite of parathion:

1) *An EDTA-stable paraoxonase*, located in Cohn-Fraction V. The activities of this enzyme, as examined with the spectrophotometric method of Krisch (1968) in all European, Mongoloid, Negroid, Aborigine and other samples studied were similarly relatively low (between 0 and 70 U/l), and demonstrated an unimodal distribution.

2) *An EDTA-labile paraoxonase*, located in Cohn-Fraction IV-1. The EDTA-labile serum paraoxonase in Europeans shows a polymorphism governed by two alleles. The first allele has a gene frequency $p_{low}$ of 0.67–0.78, resulting in the manifestation of a low activity group in homozygotes. On average 53% of all Europeans can be included in this group. A second allele (gene frequency $q_{high} = 0.22–0.33$) was found in typical European distributions resulting in the manifestation of a second heterozygotic and a third homozygotic group with medium resp. high activities. The Hardy–Weinberg rule for a two-allele model is valid only for this typical European (Caucasian) distribution.

While the Hardy–Weinberg rule for a two-allele model was exactly true for Caucasians, this was not the case for the aforementioned Negroid and Mongoloid samples where the Hardy–Weinberg rule neither for a two- nor a three-allele model could be established.

Beginning in Europe the frequency of the low activity group decreases steadily which is possibly caused by migration and mixing of the populations.

From the results of our family investigations in Vietnamese, Senegalese and Nigerian samples we believe that the largest part of Negroid and Mongoloid distribution is controlled by multiple allelomorphs.

In some of the studied samples we could not detect the low activity group.

None of these paraoxonases showed either age-related changes in activity or sex-dependent activity differences.

Individuals with high serum paraoxonase activity may be better protected against the toxic effects of parathion. This means that more than 50% of the Europeans (Caucasians) have a higher risk than non-Europeans.

## Acknowledgement

We wish to thank the Volkswagen Foundation for their generous material and financial support to this research.

## References

1. G. Schrader, *Die Entwicklung neuer insectizider Phosphorsäure-Ester* (Verlag Chemie, Weinheim, 1963).

2. R. E. Poore and R. A. Neal, *Toxicol. Appl. Pharmacol.* **23**, 759 (1972).

3. W. N. Aldrige, *Biochem. J.* **53**, 110 (1953).

4. W. N. Aldrige, *Biochem. J.* **53**, 117 (1953).

5. E. G. Erdös, C. R. Debay and M. P. Westerman, *Nature* **184**, 430 (1959).

6. E. G. Erdös, C. R. Debay and M. P. Westerman, *Biochem. Pharmacol.* **5**, 173 (1960).

7. E. G. Erdös and L. E. Boggs, *Nature* **190**, 716 (1961).

8. M. Skrinjaric-Spoljar and E. Reiner, *Biochem. Biophys. Acta* **165**, 289 (1968).

9. K. Krisch, *Z. Klin. Chem. Klin. Biochem.* **6**, 41 (1968).

10. M. Geldmacher-von Mallinckrodt, M. Rabast and H. H. Lindorf, *Arch. Toxikol.* **25**, 223 (1969).

11. M. Geldmacher-von Mallinckrodt, H. H. Lindorf, M. Petenyi, M. Flügel, T. Fischer and T. Hiller, *Humangenetik* **17**, 331 (1973).

12. M. Geldmacher-von Mallinckrodt, G. Hommel and J. Dumbach, *Hum. Genet.* **50**, 331 (1979).

13. R. Zech and K. Zuercher, *Comp. Biochem. Physiol. (B)* **48**, 427 (1974).

14. M. Geldmacher-von Mallinckrodt, T. L. Diepgen, C. Duhme and G. Hommel, *Amer. J. Phys. Anthropol.* **62**, 235 (1983).

15. J. R. Playfer, L. C. Eze, M. F. Bullen and D. A. P. Evans, *J. Med. Genet.* **13**, 337 (1976).

16. H. Eiberg and J. Mohr, *Ann. Hum. Genet.* **45**, 323 (1981).

17. G. Carro-Ciampi, D. Kadar and W. Kalow, *Can. J. Physiol. Pharmacol.* **59**, 904 (1981).

18. G. Carro-Ciampi, S. Gray and W. Kalow, *Can. J. Physiol. Pharmacol.* **61**, 336 (1983).

19. M. Geldmacher-von Mallinckrodt, M. Petenyi, H. Metzner, H. Burgis, B. Dietzel, H. Nirschl and O. Renner, *Z. Physiol. Chem.* **354**, 337 (1973).

20. M. Geldmacher-von Mallinckrodt, W. Baumgartner, M. Petenyi, H. Burgis, H. H. Lindorf and H. Metzner, *Z. Physiol. Chem.* **353**, 217 (1972).
21. H. W. Eckerson and B. N. La Du, *Drug. Metab. Dispos.* **12**, 57 (1984).
22. G. Hommel, *Biometr. J.* **20**, 371 (1978).
23. G. Hommel, C. Duhme and T. L. Diepgen, *Geniter, Interner Arbeitsbericht Nr. 180* (RRZE, Erlangen, 1983).

# The Interethnic Differences of the Human Serum Paraoxonase Polymorphism Analysed by a Quantitative and a Qualitative Method[†]

## T. L. DIEPGEN

*Institut für Medizinische Statistik und Dokumentation der Universität Erlangen, F.R.G.*

## M. GELDMACHER-VON MALLINCKRODT

*Institut für Rechtsmedizin der Universität Erlangen, F.R.G.*

and

## H. W. GOEDDE

*Institut für Humangenetik der Universität Hamburg, F.R.G.*

(*Received June 27, 1986; in final form August 1, 1986*)

The polymorphism of the human serum paraoxonase[1] was analyzed by two distinguished methods in six different ethnic groups (Caucasians, Mongoloids, Negroids), using (1) the Computer Method[2] and (2) the Carro-Ciampi Method[3,4]. Analysis of the response of the enzyme activities to salts resulting in low and high activity ratios.

*Comparison of the results:*

In Caucasians we distinguished three phenotypes by the Computer method. The polymorphism was governed by two alleles. The Hardy–Weinberg rule for a two-allele

---

†Presented at the IAEAC Workshop on Chemistry and Fate of Organophosphorus Compounds, Free University, Amsterdam, June 18–20, 1986.

This article was first published in *Toxicological and Environmental Chemistry*, Volume 14, Numbers 1 + 2 (1987).

C

model was valid. Individuals belonging to the homozygotic group with low activity had a low activity ratio (Carro-Ciampi method). With both methods a frequency between 57% and 61% was observed for this group. Individuals with medium and high activity had a high activity ratio.

In Negroids and Mongoloids samples we found (by the Computer method) a low activity group (Ghanaians 9.6%, Jamaicans 13.6%, Indonesians 6.7%, Koreans 19.6%). The Hardy–Weinberg rule for a two- or three-allele model was not valid. Individuals belonging to the low activity group had a low activity ratio, all individuals with higher activity a high activity ratio (Carro-Ciampi method).

Our results suggest that the members of the low activity group in the three races are homozygote for an identical allele.

KEY WORDS:    Paraoxonase, ethnic groups, polymorphism, serum samples, enzyme activity, spectrophotometry.


## INTRODUCTION

The activity of the enzyme serum paraoxonase is polymorphically distributed in many populations with strong interethnic differences.[1,3–12]

The polymorphism of the paraoxon-hydrolysing activity of serum in Europeans is governed by a two-allele model of codominant inheritance resulting in three phenotypes with low, medium and high activity (gene frequencies $p_{low} = 0.67$ to $0.78$; $q_{high} = 0.22$ to $0.33$).[1] While the Hardy–Weinberg rule for a two-allele model was true for Caucasians, this was not the case for the negroid and mongoloid samples studied, where the Hardy–Weinberg rule for neither a two-allele nor a three-allele model could be established,[1] though most of them showed the low activity group. In Figure 1 it can be seen that the frequencies of the low activity group tend to decrease moving away from Europe.[13]

These measurements were performed using the kinetic test of Krisch.[5] The polymorphical distribution of the enzyme activity was separated by the computer program GENITER[2] into groups on the basis of a mathematical model.[14]

Eckerson *et al.*[11] developed a method for identifying two human serum paraoxonase phenotypes based upon the effect of NaCl on the paraoxonase activity.

In view of that finding, Carro-Ciampi *et al.*[3,4] described the paraoxonase phenotype distribution in different Canadian popu-

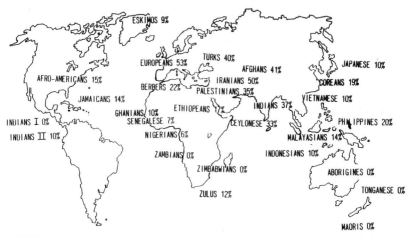

FIGURE 1 Percentages of phenotypes with low serum paraoxonase activity in random samples taken from different ethnic groups (Diepgen and Geldmacher-v. Mallinckrodt)[13].

lations (Caucasians $n = 82$; Indians $n = 57$; Inuits $n = 67$). Using a direct spectrophotometric method, three different determinations of paraoxon hydrolysis rates were carried out for each serum sample:

a) buffer alone,

b) buffer $+ CaCl_2$,

c) buffer $+ CaCl_2 + NaCl$.

The values of the ratio (b)/(c) designating enzyme activation by $Na^+$ in the presence of $Ca^{++}$, were clearly bimodally distributed.

The goal of our study was the further exploration of the genetic polymorphism of the human serum paraoxonase in Caucasians, Mongoloids and Negroids using both methods and comparing their results.

## METHODS

### Subjects

Human sera were obtained from unrelated volunteers of different ethnic groups: 130 West Germans, 96 Italians, 120 Indonesians, 106 South Koreans, 93 Ghanaians and 129 Jamaicans.

All the test subjects were healthy and to their own knowledge did not suffer from any metabolic diseases or liver disorders.

## Chemical and statistical methods

*Computer method*   The paraoxonase activity was measured according to Krisch.[5] On the basis of a mathematical model for separation of a population into groups according to their enzyme activities[14] the computer-program GENITER[2] described the empirical polymorphic distribution of the paraoxonase activity by several normal distributed groups. It estimated relative frequencies $q_i$ of the presumed groups and calculated the means $\mu_i$ and variances $\sigma^2$ by an iterative procedure. Assuming a two-allele model of codominant inheritance we distinguished between three normal distributed phenotypes in the trimodal distribution of the enzyme activity. The chi-square test was applied to test the validity of the Hardy–Weinberg rule.[15]

*Carro-Ciampi method*   The hydrolysis rates of paraoxon were measured under three different conditions:

  a) in glycine-NaOH buffer,

  b) in glycine-NaOH buffer $+ 5 \times 10^{-3}$ M $CaCl_2$,

  c) in glycine-NaOH buffer $+ 5 \times 10^{-3}$ M $CaCl_2 + 5 \times 10^{-1}$ M NaCl.

The phenotyping ratio (c)/(b) was determined and their distribution plotted.

*Comparison of the two methods*   For the data analysis the values of paraoxonase activity at pH 11.2 and of (c)/(b) ratio were plotted on separate frequency distribution histograms, two for each ethnic group. The existence of a correlation between paraoxonase activity at pH 11.2 and phenotyping ratio (c)/(b) or between (b) and (c) was analyzed by linear regression. The chi-square test was applied to analyze the cross-classified low and high activity clusters of both methods.

## RESULTS

The percentage frequencies $q_i$, means $x_i$ and standard deviations $s_i$

TABLE I

Iteration results of percentage frequencies $q_i$, means $\bar{x}_i$ and standard deviations $s_i$ of the three phenotypes $(i = 1, 2, 3)$ of the serum paraoxonase activity $(pH = 11.2)$ in the listed samples assuming a two-allele model.

| Europeans | 130 Germans ($p > 0.50$) | | | 96 Italians ($p > 0.60$) | | |
|---|---|---|---|---|---|---|
| | $i = 1$ | $i = 2$ | $i = 3$ | $i = 1$ | $i = 2$ | $i = 3$ |
| $q_i$ (%) | 57.3 | 35.5 | 7.7 | 60.7 | 33.3 | 6.0 |
| $x_i$ ($U/1$) | 64.1 | 213.3 | 317.0 | 60.3 | 202.2 | 362.0 |
| $s_i$ ($U/1$) | 15.5 | 45.3 | 72.7 | 11.2 | 47.7 | 27.8 |

| Mongoloids | 120 Indonesians ($p < 0.05$) | | | 106 Koreans ($p < 0.0005$) | | |
|---|---|---|---|---|---|---|
| | $i = 1$ | $i = 2$ | $i = 3$ | $i = 1$ | $i = 2$ | $i = 3$ |
| $q_i$ (%) | 6.7 | 24.6 | 68.7 | 19.6 | 21.6 | 58.8 |
| $x_i$ ($U/1$) | 78.5 | 164.7 | 256.4 | 62.3 | 160.4 | 269.4 |
| $s_i$ ($U/1$) | 4.2 | 21.3 | 84.8 | 14.4 | 23.1 | 68.5 |

| Negroids | 93 Ghanaians ($p < 0.05$) | | | 129 Jamaicans ($p < 0.005$) | | |
|---|---|---|---|---|---|---|
| | $i = 1$ | $i = 2$ | $i = 3$ | $i = 1$ | $i = 2$ | $i = 3$ |
| $q_i$ (%) | 9.6 | 28.9 | 61.5 | 13.6 | 29.7 | 56.7 |
| $x_i$ ($U/1$) | 44.03 | 157.1 | 276.5 | 48.9 | 160.6 | 247.3 |
| $s_i$ ($U/1$) | 16.0 | 23.9 | 49.3 | 20.5 | 32.2 | 75.4 |

for the three phenotypes $(i = 1, 2, 3)$ of the paraoxon hydrolyzing activity estimated by the computer program GENITER is seen in Table I. The vaiidity of the Hardy–Weinberg rule cannot be rejected for the Caucasians ($p > 0.50$; $p > 0.60$) and must be refuted for the Mongoloides ($p < 0.05$, $p < 0.0005$) and the Negroids ($p < 0.05$, $p < 0.005$).

Figure 2 shows the distribution of the serum paraoxonase activity and of the phenotyping ratio (c)/(b) in our six different ethnic groups. The phenotyping ratio (c)/(b) shows a bimodal distribution for all six different ethnic samples. A (c)/(b) ratio $< 1.20$ represents in all cases the "low-activity" phenotype and a (c)/(b) ratio $> 1.20$ the "high-activity" phenotype, i.e. the homozygous and the heterozygous high genotype.

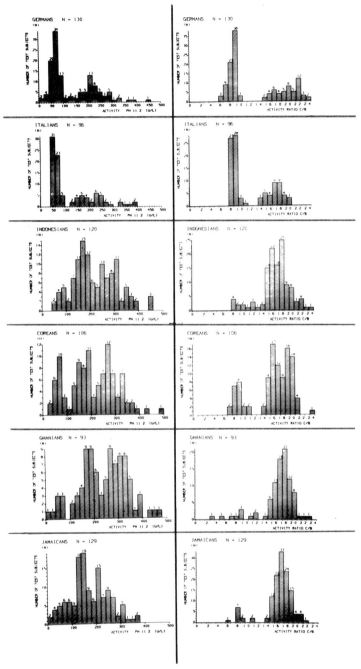

FIGURE 2 Frequency distribution histograms of serum paraoxonase activity (pH = 11.2) and of serum paraoxonase activity ratio (c)/(b) (pH = 10.0) in different ethnic samples.

52

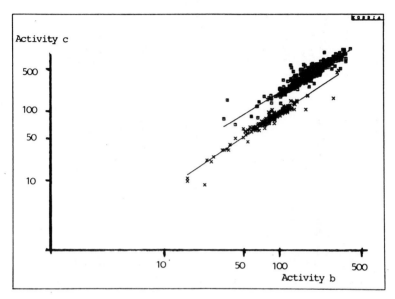

FIGURE 3 Scattergram of paraoxon-hydrolysing activities in human sera. Determinations were made in (b) glycine NaOH buffer plus $CaCl_2$ and in (c) as in (b) plus NaCl. × (c)/(b) < 1.2. □ (c)/(b) > 1.2.

A clearly apparent correlation between activity (b) and (c), plotted in pairs on a logarithmic scale, for all samples is shown in Figure 3 for the (c)/(b) ratio < 1.20 as well as > 1.20. The data also show, that there is a strong correlation between activity (b) and (c) if a distinction between (c)/(b) values higher and lower than 1.2 is made simultaneously. This is true for all different ethnic groups studied.

In Table II we compare the relative frequencies of the low activity as calculated by GENITER for different ethnic groups with the respective percentages of the low (c)/(b) ratio (< 1.2) determined according to the Carro-Ciampi method.

The strong dependence between clusters of low and high activity separated by the computer method on the one hand and by the enzyme response to salts (cut off (c)/(b) = 1.20) on the other hand shows the chi-square test of all cross classified data ($p \ll 0.001$). The good correlation of the results analysed by both the methods is shown in the scattergram Figure 4 for all test samples of the six different ethnic groups.

T. L. DIEPGEN *ET AL.*

## TABLE II

Frequencies of the genotypes of the serum paraoxonase activity in six different ethnic groups as estimated by GENITER computer program. The percentages of low and high serum paraoxonase activity according to the Carro-Ciampi method are also shown.

| | GENITER | | | Carro-Ciampi method | |
|---|---|---|---|---|---|
| | homo-zygous low | hetero-zygous high | homo-zygous high | low ratio (c)/(b) < 1.20 | high ratio (c)/(b) > 1.20 |
| Germans ($n=130$) | 57.3 | 35.5 | 7.7 | 56.9 | 43.1 |
| Italians ($n=96$) | 60.7 | 33.3 | 6.0 | 61.5 | 38.5 |
| Indonesians ($n=120$) | 6.7 | 24.6 | 68.7 | 7.5 | 92.5 |
| Koreans ($n=106$) | 19.6 | 21.7 | 58.7 | 19.8 | 80.2 |
| Ghanaians ($n=93$) | 9.6 | 28.9 | 61.5 | 9.7 | 90.3 |
| Jamaicans ($n=129$) | 13.6 | 29.7 | 56.7 | 9.3 | 90.7 |

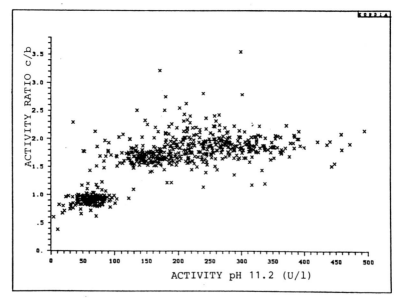

FIGURE 4 Scattergram between paraoxon-hydrolysing activity pH = 11.2 and activity ratio (c)/(b).

We compared the percentage frequencies $q_1$ estimated by GENITER with the "low-activity" phenotype represented by the (c)/(b) ratio $< 1.20$ in Table II. In Caucasians about 60% belong to that group in contrast to Mongoloids and Negroids, where we find a percentage frequency between 6.7% to 19.6%.

## DISCUSSION

The paraoxonase hydrolysing activity of human serum in Caucasians is clearly trimodally and in most Mongoloids and Negroids bi- and/or trimodally distributed.[1] We separated each population into high and low paraoxonase activity using the response to NaCl and $CaCl_2$ (Carro-Ciampi et al.)[3,4] and the GENITER computer program.[2] We found a good correlation between the results of both methods.

Carro-Ciampi et al.[3,4] applied their method to Caucasians as well as Indians and Inuits, who are both Mongoloids. They found that an activity ratio (c)/(b) = 1.5 is suitable for separation of high and low activity of human serum paraoxonase. Using typical Mongoloid samples (Indonesians and Koreans) we could show that both the Carro-Ciampi method and our GENITER computer program arrange the same individuals in the same activity group. The same applies for a typical Negroid (Ghanaians) and a mixed population (Jamaicans).

Thus we could show that the Carro-Ciampi method finds qualitative differences between human serum paraoxonase for all three major races. The technique is suitable to classify an individual as a member of the low or high activity group. In contrast to Carro-Ciampi et al.[3,4] we found in our 675 sera of different ethnic groups that there is a cut off at a (c)/(b) ratio of 1.2 and not 1.5. Both the Carro-Ciampi and our computer method show that more than 50% of the Caucasians and less than 20% of the Mongoloids and Negroids belong to the low paraoxonase activity group.

Even though the validity of the Hardy–Weinberg rule assuming a two- or three-allele model must be rejected in the Mongoloids and Negroids using the computer method, the response to salts ((c)/(b) $< 1.2$) is identically in all groups with low activity. This would be in accordance with the presumed identity of the low activity allele in all ethnic groups.

In addition to that the GENITER computer program is suitable to show that under assumption of a two-allele model and codominant inheritance in Caucasians there are homozygous and heterozygous members for the high activity group. The computer program can also estimate gene frequencies and for those the validity of the Hardy–Weinberg rule can be tested. The same procedure can be applied for the estimation of distribution parameters of polymorphic distributions if a three-allele model is assumed.

## References

1. M. Geldmacher-v. Mallinckrodt, T. L. Diepgen, C. Duhme and G. Hommel, *Amer. J. Phys. Anthropol.* **62**, 235 (1983).
2. G. Hommel, C. Duhme and T. L. Diepgen, *Interner Arbeitsbericht Nr. 180* (RRZE, Erlangen, 1983).
3. G. Carro-Ciampi, D. Kadar and W. Kalow, *Can. J. Pharmacol.* **59**, 904 (1981).
4. G. Carro-Ciampi, S. Gray and W. Kalow, *Can. J. Physiol. Pharmacol.* **61**, 336 (1983).
5. K. Kirsch, *Z. Klin. Chem. Klin. Biochem.* **6**, 41 (1968).
6. R. Zech and K. Zuercher, *Comp. Biochem. Physiol.* (*B*) **48**, 427 (1974).
7. M. Geldmacher-v. Mallinckrodt, U. Rabast and H. H. Lindorf, *Arch. Toxicol.* **25**, 223 (1969).
8. M. Geldmacher-v. Mallinckrodt, G. Hommel and J. Dumbach, *Hum. Genet.* **50**, 313 (1979).
9. J. R. Playfer, L. C. Eze, M. F. Bullen and D. A. P. Evans, *J. Med. Genet.* **13**, 337 (1976).
10. H. Eiberg and J. Mohr, *Ann. Hum. Genet.* **45**, 323 (1981).
11. H. W. Eckerson, J. Romson, C. Wyte and B. N. La Du, *Am. J. Hum. Genet.* **35**, 214 (1983).
12. R. F. Mueller, S. Hornung, C. E. Furlong, J. Anderson, E. R. Giblett and A. G. Motulsky, *Am. J. Hum. Genet.* **35**, 393 (1983).
13. T. L. Diepgen and M. Geldmacher-v. Mallinckrodt, *Arch. Toxicol. Suppl.*, 154 (1986).
14. G. Hommel, *Biometr. J.* **20**, 371 (1978).
15. H. Cramer, *Mathematical Methods of Statistics* (Princeton University Press, Princeton, 1946), Ch. 30, pp. 424–434.

# The Metabolites of Organophosphorus Pesticides in Urine as an Indicator of Occupational Exposure†

Ž. VASILIĆ, V. DREVENKAR, Z. FRÖBE, B. ŠTENGL and B. TKALČEVIĆ

*Institute for Medical Research and Occupational Health, 41000 Zagreb, Yugoslavia*

(*Received July 22, 1986; in final form August 10, 1986*)

The differences in the degree in workers' exposure to organophosphorus pesticides during the spraying of an apple-orchard were assessed from the urinary content of metabolites: dimethyl phosphorothiolate potassium salt (DMPThK) for Demeton-S-methyl and dimethyl phosphorothionate and phosphorodithioate potassium salts (DMTPK and DMDTPK) for Azinphos-methyl and Methidathion. The highest median concentrations of the metabolites in the urine samples collected after two to three days of work with the pesticides were determined in the mixers preparing pesticide solutions: DMPThK 83 ng cm$^{-3}$ ($N=7$) after exposure to Demeton-S-methyl; DMTPK 2040 and DMDTPK $<20$ ng cm$^{-3}$ ($N=7$) after exposure to Azinphos-methyl; DMTPK 501 and DMDTPK 88 ng cm$^{-3}$ ($N=6$) after exposure to Methidathion. The applicators (sprayers) were the second most exposed group: DMPThK 30 ng cm$^{-3}$ ($N=6$) after exposure to Demeton-S-methyl; DMTPK 433 and DMDTPK $<20$ ng cm$^{-3}$ ($N=7$) after exposure to Azinphos-methyl; DMTPK 45 and DMDTPK $<20$ ng cm$^{-3}$ ($N=16$) after exposure to Methidathion. The median concentrations in the other orchard employees ($N=5$–8) were always $\leq 50$ ng cm$^{-3}$.

†Presented 18–20 June 1986 at the IAEAC Workshop on Chemistry and Fate of Organophosphorus Compounds, Free University, Amsterdam.

This article was first published in *Toxicological and Environmental Chemistry*, Volume 14, Numbers 1 + 2 (1987).

Mean decrease/increase in pre-exposure cholinesterase activity in whole blood and plasma (to 20%) gave no indication of different organophosphorus pesticide absorption in different groups of workers. The concentration of specific metabolites is an early, direct, more reliable and more sensitive indicator of subtle differences in occupational exposure which should initiate well-timed and more efficient measures for the protection of workers' health.

KEY WORDS:    Organophosphorus pesticides, urinary metabolites, blood cholinesterase activity, occupational exposure, gas chromatography.

## INTRODUCTION

Occupational exposure to organophosphorus (OP) pesticides can be monitored indirectly by measuring cholinesterase activity in plasma, red blood cells and whole blood[1-4] and directly by determining characteristic pesticide metabolites in blood[5,6] and urine.[1,7-15]

Cholinesterase is a nonspecific indicator of exposure to OP pesticides. It is influenced also by absorption of other inhibitors as well as by additional impacts not connected with exposure.[4,9] The monitoring of changes in baseline cholinesterase activity (pre-exposure values) was proposed as the most effective method for evaluating safety measures taken to protect agricultural workers from exposure to OP pesticides and carbamates.[16] However, a progressive decrease in cholinesterase activity as a result of multiple exposures was rarely noticed. This circumstance was explained by an accelerated rate of enzyme regeneration in the body if small amounts of pesticides were absorbed each day.

The metabolic degradation of OP pesticides which are mostly triesters of phosphoric, thiophosphoric and dithiophosphoric acids, results in humans in urinary excretion of alkali metal salts of the corresponding dialkyl esters. Dialkyl phosphates originate from all three types of triesters, dialkyl phosphorothioates from the last two types and dialkyl phosphorodithioates only from the last one. The determination of these metabolites in urine is proposed as a sensitive and specific indicator of exposure enabling the detection of absorbed pesticide amounts smaller than those depressing cholinesterase activity. The presence of specific metabolites in urine proved to be a reliable sign of exposure to one[8-10,15,17] or to more OP pesticides.[7,11-14,16] Their amount may serve as an index of pesticide absorption although, alone, it does not offer any evidence of the toxic effect.

The studies reported in this work originate from the experimental observation based on routine monitoring of cholinesterase activity that this parameter is not sensitive enough to indicate subtle differences in exposure to OP pesticides of different groups of agricultural workers. To find out whether some workplaces are especially at risk because of the use of inadequate technological procedure or deficient protective devices we measured urinary excretion of specific dialkylphosphorus metabolites in workers employed in an apple orchard during treatment with OP pesticides Demeton-S-methyl, Azinphos-methyl and Methidathion. The results were compared with the data on cholinesterase activity in whole blood and plasma.

## MATERIALS AND METHODS

### Standards and reagents

Standard solutions of OP pesticide metabolites in deionized water were prepared with 0,0-dimethyl phosphorothionate potassium salt (DMTPK) Lot. No. A 724, 0,0-diethyl phosphorothionate potassium salt (DETPK) Lot. No. A 468, 0,0-dimethyl phosphorodithioate potassium salt (DMDTPK) Lot. No. A 858 and 0,0-diethyl phosphorodithioate potassium salt (DEDTPK) Lot. No. A 459, all supplied by the United States Environmental Protection Agency Repository, Research Triangle Park, NC, U.S.A.

Diazomethane ethereal solution in a concentration of $10 \pm 5$ mg $CH_2N_2$ cm$^{-3}$ was prepared from N-methyl-N-nitroso-$p$-toluene-sulphonamide purchased from Merck-Schuchardt, Germany, by means of common distillation procedure.[18]

Diethyl ether p.a., chromatography grade purity, calcium hydroxide p.a. and hydrochloric acid p.a. were products of "Kemika", Yugoslavia. Diethyl ether was redistilled before use.

### Apparatus

A Varian Aerograph Series 2800 gas chromatograph equipped with an alkali flame ionization detector ($Rb_2SO_4$) was used. A 1.8 m × 2 mm i.d. gas chromatographic column was packed with 4%

SE-30+6% OV-210 on Gas Chrom Q 0.16–0.20 mm; the first 10 cm of the column were packed with 10% Carbowax 20M on Chromosorb W/NAW 0.16–0.20 mm. The operating temperatures of the column, injector and detector were 95°C, 215°C and 230°C respectively; nitrogen (carrier gas), air and hydrogen flow rates were 30, $235 \pm 10$ and $35 \pm 3$ cm$^3$ min$^{-1}$ respectively.

## Quantitative determination of dialkylphosphorus metabolites of OP pesticides in urine

The urine samples taken from occupationally exposed workers and controls were treated according to the method reported previously.[10] To a 5-cm$^3$ aliquot of the sample 0.5 cm$^3$ of deionized water was added. Inorganic phosphates were removed by adding 0.1 g Ca(OH)$_2$, mixing for 1 min on a Vortex mixer and centrifuging at 424 G for 3 min. One cubic centimeter of the supernatant was transferred into a test tube with a ground stopper and 4 cm$^3$ of diethyl ether and 1 cm$^3$ of 6 mol dm$^{-3}$ HCl were added. After 1 min of vortexing, 2 cm$^3$ of the ethereal layer was separated and alkylated by addition of 1 cm$^3$ of diazomethane solution. The sample was left to stand 10 min and the solution was evaporated to 1 cm$^3$ under a stream of nitrogen, so that the excess of diazomethane was also removed.

Standards for quantitative gas chromatographic evaluation of OP pesticide metabolites were prepared by adding 0.06—0.5 cm$^3$ of a mixture of DMTPK, DETPK, DMDTPK and DEDTPK dissolved in deionized water to 5 cm$^3$ urine samples taken from non-exposed persons. Deionized water was added to make the final volume of 5.5 cm$^3$. The concentrations of single metabolites in water solution were from 12.78 to 17.08 $\mu$g cm$^{-3}$. The final concentrations in urine standards were in the range from 150 to 1700 ng cm$^{-3}$. Further on the urine standards were treated in the same way as the urine samples of exposed workers and controls.

## Determination of cholinesterase activity

Whole blood and plasma cholinesterase activities were measured spectrophotometrically by means of Ellman's method.[19]

## Subjects

The workers were grouped according to their job and level of exposure to pesticides. The mixers mixing and loading pesticide solutions always made the first (I) and applicators (sprayers) the second (II) group. The composition of the third (III) group varied. It consisted of other persons who were not directly involved in the preparation and application of pesticides, i.e. of mechanics servicing and cleaning the tractors and instruments used for the orchard spraying, field workers and a few other employees.

The mixers never used a closed system for transferring the concentrated liquid or powdered pesticides from their original containers to mix and spray tanks. The mixers and applicators were obliged to wear protective clothing which reduced but did not completely eliminate the possibility of respiratory and dermal exposure.

The exposure to Demeton-S-methyl [S-(ethylthio)ethyl dimethyl phosphorothiolate] was estimated by analysis of urine samples from seven mixers (I), six applicators (II) as well as from a group of seven persons comprising three mechanics, one store-house worker, one assistant manager, one housekeeper and one manager (III). The samples were taken one day before work and immediately after the end of the orchard treatment on the third day. All workers worked eight hours a day.

The treatment of orchard with Azinphos-methyl [S-(3,4-dihydro-4-oxobenzo |d|-|1,2,3|-triazin-3-ylmethyl dimethyl phosphorothiolothionate] lasted two days. Urine samples were collected from 7–8 mixers (I), 7–16 applicators (II), and 6–7 mechanics (III) before work on the first day, at the end of work on the second day, and again 12–18 hours after the work was over. All the workers worked one 8-hours shift on the first day and most of them worked an 8-hour shift on the second day; three mixers and two applicators worked 14 hours on the second day.

The concentration of characteristic metabolites during the three-day orchard treatment with Methidathion [S-(2,3-dihydro-5-methoxy-2-oxo-1,3,4-thiadiazol-3-ylmethyl) dimethyl phosphoro-thiolothionate] was measured in the urine samples collected from six mixers (I), 11–16 applicators (II) and eight field workers (III) one day before they started working and on the third day immediately after they finished work. The workers worked eight hours a day. One mixer and one applicator worked each day in two shifts, doing their normal work during the first and simply being present on the campus during the second shift.

Concurrently with the analysis of urine samples taken from occupationally exposed persons, 11 urine samples collected from non-exposed persons were analysed for DMTPK, DMDTPK, DETPK and DEDTPK.

The results of analysis were not corrected for individual differences in urinary excretion by any adjustment of metabolite concentrations. Collection of more reliable 24-hour or shorter timed samples from exposed workers was impracticable.

The inhibition of cholinesterase activity in whole blood and plasma of exposed workers was determined by comparing individual's pre-exposure cholinesterase activity measured immediately before the beginning of orchard treatment with one of the pesticides and the cholinesterase activity in the blood sample taken from the same individual at the end of the third day of work with Demeton-S-methyl and one day after the end of work with Azinphos-methyl or Methidathion.

The individual variations in pre-exposure cholinesterase activity were not determined because it was not possible to collect blood samples more than once before work with OP pesticides.

## RESULTS AND DISCUSSION

### Analysis of OP pesticide metabolites in the urine of exposed workers

The urinary excretion of alkali metal salts of dimethyl phosphoro-dithioate, phosphorothionate and phosphate takes place as a result of the metabolic degradation of Methidathion and Azinphos-methyl. Similarly, Demeton-S-methyl, an 0,0,S-substituted phosphorothiolate, is metabolized to dimethyl phosphorothiolate and dimethyl phosphate. For the control of workers occupationally exposed to Methidathion or Azinphos-methyl the first two characteristic metabolites were determined in urine. The exposure to Demeton-S-methyl was monitored by urinary analysis of dimethyl phosphorothiolate. Dimethyl phosphate was omitted because of interferences of small amounts of inorganic phosphates which remained in urine despite precipitation with calcium hydroxide.[7] In the analytical procedure applied inorganic phosphate and dimethyl phosphate were both transformed to trimethyl phosphate by methylation before gas chromatographic analysis.

The extraction recoveries of investigated metabolites ranged from 80 to 95%. To minimize possible errors caused by nonquantitative recoveries, the standards were prepared by addition of potassium salts of metabolites to the urine of non-exposed persons and treated in the same way as the urine samples of exposed persons.

In default of a dimethyl phosphorothiolate (DMPThK) standard the concentration of this metabolite in workers exposed to Demeton-S-methyl was calculated using the standards prepared from DMTPK. During methylation owing to the thiono–thiolo iso-merization DMTPK yields both, O-methyl and S-methyl isomers. In a protic solvent such as methanol the thiolo isomer is formed with a yield of over 90%[20] and in the ethereal extract obtained by the procedure used in this work with a 93% yield.

In non-exposed persons the reported urinary concentrations of dialkylphosphorus metabolites of OP pesticides varied between 20 and 40 ng cm$^{-3}$.[12] In this work DMTPK was detected in four and DMDTPK in one out of 11 urine samples in the concentration range from <20 to 60 ng cm$^{-3}$. Analogous diethyl derivatives were not detected at all.

The results of DMPThK analysis in urine before and after workers' exposure to Demeton-S-methyl are presented in Table I. The pre-exposure concentrations were in all groups within the range determined for non-exposed persons.[12] After exposure of three days the highest median value and the highest concentration in general were determined in urine samples collected from mixers, followed by the group of applicators and the group of other workers. The concentrations above 100 ng cm$^{-3}$, which were related to exposure,[12] were found in three mixers and two applicators and concentrations between 40 and 100 ng cm$^{-3}$, indicating a slight exposure,[12] in one mixer, one applicator and one worker from group III. Despite the not very high urinary concentrations of DMPThK in all groups of workers, most of which were below 260 ng cm$^{-3}$, the mixers seemed to be the most exposed group.

TABLE I

Concentration of DMPThK metabolite of Demeton-S-methyl in the
urine of exposed workers: (A) one day before the beginning of work;
(B) at the end of the third day of work with Demeton-S-methyl.

| Sample | Number of workers | Geometrical mean value $(\text{ng cm}^{-3})$ | Median $(\text{ng cm}^{-3})$ | Range $(\text{ng cm}^{-3})$ |
|---|---|---|---|---|
| *Mixers* | | | | |
| A | 7 | 30 | 0 | 0– 30 |
| B | 7 | 121 | 83 | 0–822 |
| *Applicators* | | | | |
| A | 6 | 32 | 30 | 0– 42 |
| B | 6 | 91 | 30 | 0–208 |
| *Other workers*[a] | | | | |
| A | 7 | 53 | 0 | 0– 53 |
| B | 7 | 42 | 30 | 0–100 |

[a]Three mechanics, one store-house worker, one assistant manager, one manager, and one housekeeper.

0 = Values lower than detection limit ( $< 30 \text{ ng cm}^{-3}$).

This was confirmed by analysis of DMTPK and DMDTPK in urine samples of mixers, applicators and mechanics collected before and after application of Azinphos-methyl in the orchard. The results are summarized in Table II. Among altogether 25 urine samples collected on the first day, before work, the urines of one mixer and three applicators were found to contain traces of DMTPK only. However, after two days of exposure DMTPK concentrations increased significantly in all mixers and in six out of seven applicators. The median DMTPK value in mixers was approximately five times higher than the corresponding median value for applicators. Three mixers who worked two successive shifts on the second day, also had the highest DMTPK concentrations in urine: 3238, 3439 and $5556 \text{ ng cm}^{-3}$. An analogous difference in DMTPK urinary level in the group of applicators was not observed despite longer working hours of two applicators.

In the urine samples collected 12–18 hours after work with Azinphos-methyl DMTPK was still present in all mixers and in 14 out of 16 applicators, but with approximately four times lower

TABLE II

Concentration of DMTPK and DMDTPK metabolites of Azinphos-methyl in the urine of exposed workers: (a) on the first day before work; (b) on the second day immediately after work; (c) 12–18 hours after the end of work with Azinphos-methyl.

| Sample | Number of workers | Geometrical mean value $(\text{ng cm}^{-3})$ | | Median $(\text{ng cm}^{-3})$ | | Range $(\text{ng cm}^{-3})$ | |
|---|---|---|---|---|---|---|---|
| | | DMTPK | DMDTPK | DMTPK | DMDTPK | DMTPK | DMDTPK |
| *Mixers* | | | | | | | |
| a | 7 | 33 | 0 | 0 | 0 | 0– 33 | 0 |
| b | 7 | 2111 | 250 | 2040 | 0 | 975–5556 | 0–700 |
| c | 8 | 582 | 147 | 569 | 0 | 138–2781 | 0–248 |
| *Applicators* | | | | | | | |
| a | 12 | 54 | 0 | 0 | 0 | 0– 96 | 0 |
| b | 7 | 292 | 0 | 433 | 0 | 0– 945 | 0 |
| c | 16 | 109 | 0 | 102 | 0 | 0–1110 | 0 |
| *Mechanics* | | | | | | | |
| a | 5 | 0 | 0 | 0 | 0 | 0 | 0 |
| b | 5 | 32 | 0 | 24 | 0 | 0– 47 | 0 |
| c | 6 | 44 | 0 | 28 | 0 | 0– 238 | 0 |

$0 =$ Values lower than detection limit ($< 20$ ng cm$^{-3}$).

median values. The ratio of median values between these two groups of workers remained 5:1 as on the day before.

Another Azinphos-methyl metabolite DMDTPK was detected only in mixers, i.e. in three out of six and in two out of eight urine samples taken immediately after work and 12–18 hours later, respectively.

Mechanics were obviously the least exposed group of workers. Only traces of DMTPK ($< 50$ ng cm$^{-3}$) were detected in their urine samples collected in two samplings after work with no more than one exception in the second sampling (238 ng cm$^{-3}$).

During the treatment of the orchard with Methidathion instead of negligibly exposed mechanics the field workers not working directly with pesticides were included as the third group. In addition to Methidathion metabolites DMTPK and DMDTPK the urinary excretion of DETPK and DEDTPK was also measured. The aim

TABLE III

Concentration of DMTPK and DMDTPK metabolites of Methidathion and of DETPK in urine samples of exposed workers: (a) one day before the beginning of work: (b) at the end of the third day of work with Methidathion.

| Metabolite | Number of workers | | Geometrical mean value (ng cm$^{-3}$) | | Median (ng cm$^{-3}$) | | Range (ng cm$^{-3}$) | |
|---|---|---|---|---|---|---|---|---|
| | a | b | a | b | a | b | a | b |
| *Mixers* | | | | | | | | |
| DMTPK | 6 | 6 | 26 | 604 | 25 | 501 | 20–46 | 173–4506 |
| DMDTPK | 6 | 6 | 21 | 91 | 10 | 88 | 0–22 | 24– 544 |
| DETPK | 6 | 6 | 35 | 48 | 0 | 41 | 0–35 | 0– 119 |
| *Applicators* | | | | | | | | |
| DMTPK | 16 | 11 | 37 | 49 | 0 | 45 | 0–87 | 20– 122 |
| DMDTPK | 16 | 11 | 20 | 63 | 0 | 0 | 0–20 | 0– 65 |
| DETPK | 16 | 11 | 20 | 39 | 0 | 24 | 0–20 | 0– 145 |
| *Field workers* | | | | | | | | |
| DMTPK | 8 | 8 | 26 | 41 | 0 | 51 | 0–38 | 20– 80 |
| DMDTPK | 8 | 8 | 20 | 28 | 0 | 26 | 0–20 | 0– 36 |
| DETPK | 8 | 8 | 0 | 88 | 0 | 63 | 0 | 0– 279 |

0 = Values lower than detection limit ( < 20 ng cm $^{-3}$).

was to check if the workers, previously or simultaneously, were exposed to esters of diethyl dithiophosphoric acid (e.g. Phosalone) which were also frequently used for protection of orchard and were stored in the orchard store-house. The results are presented in Table III. According to the urinary excretion of DMTPK and DMDTPK mixers were again the most exposed group of workers, all having traces of DMTPK even one day before beginning work with Methidathion. On the same day similar low concentrations of DMTPK were detected also in seven out of 16 applicators and in three out of eight field workers. Minimal amounts of DMDTPK were present in urine samples of three mixers, one applicator and one field worker only. At the end of three days of work with Methidathion DMTPK and DMDTPK concentrations in mixers increased up to 100 and 30 times, respectively. DMTPK was also present in the urine of all applicators but in a lower concentration range and with 11 times lower median value compared to the group of mixers. The DMDTPK concentration level, which was determined in four out of 11 applicators was only slightly different with respect to the results obtained before work with Methidathion.

The urinary excretion of Methidathion metabolites in one mixer and one applicator who worked on the first shift but were also present in the orchard during the second shift each day and in other workers from groups I and II was not different.

The lowest urinary level of DMTPK and DMDTPK after three days of exposure to Methidathion was noticed in field workers but DMTPK was present in all and DMDTPK in six out of eight urine samples.

Of two other measured OP pesticide metabolites DEDTPK was not detected in any of the urine samples collected in the period of orchard treatment with Methidathion. However, contrary to the negligible concentrations of DETPK found only in one mixer and in one applicator before work, after three days of work with Methidathion the same metabolite was present at nearly the same concentration level in five out of six mixers, in seven out of 11 applicators and in seven out of eight field workers; in one mixer, one applicator and three field workers the DETPK concentration was above $100 \, \mathrm{ng \, cm^{-3}}$. The presence of DETPK in urine samples of most workers after treatment of orchard with Methidathion seemed more likely to be due to the treatment than to common exposure. Possible sources of DETPK presence in urine could be (1) the impurities in Methidathion preparation; (2) the residues of pesticides applied previously left over in improperly cleaned vessels where spraying solutions had been prepared as well as on tractors and spraying instruments; and (3) an extended stay of workers in the space where different pesticides were stored and handled.

The data on urinary excretion of Demeton-S-methyl, Azinphos-methyl and Methidathion, in different groups of workers were compared statistically by the Kruskal–Wallis test.[21] A significant difference at the 0.5% level was confirmed for DMTPK concentrations in mixers, applicators and mechanics 12–18 hours after work with Azinphos-methyl ($H = 12.4 >$ significance limit $= 10.6$) and for DMTPK and DMDTPK concentrations in mixers, applicators and field workers immediately after the three-day treatment of orchard with Methidathion ($H = 13.7$ and $9.65$ respectively $>$ significance limit $= 10.6$).

According to the same statistical test the urinary DMPThK concentrations in different groups of workers after a thre-day exposure to Demeton-S-methyl and the DETPK concentrations in mixers, applicators and field workers after exposure to Methidathion were not significantly different at the 5% level ($H = 3.60$ and 4.37 respectively < significance limit = 5.99).

## Analysis of cholinesterase activity in whole blood and plasma in exposed workers

The pre-exposure values of workers' cholinesterase activity in whole blood and plasma before the beginning of the spraying season were not known. Therefore the cholinesterase activities measured before work with Demeton-S-methyl, Azinphos-methyl or Methidathion were taken as baseline values for each worker and compared with the activities determined after the treatment of orchard when an inhibition due to pesticide application was expected.

The cholinesterase activities after workers' exposure to Demeton-S-methyl, Azinphos-methyl or Methidathion expressed as percentages of baseline values, are shown in Tables IV, V and VI, respectively. A decrease in enzyme activity, to 20%, found in most workers in all groups could not be regarded as a reliable indicator of exposure, being in the range of possible individual physiological variations. According to analysis of variance a significant difference in plasma cholinesterase activity at the 5% level was established only in different groups of workers after work with Demeton-S-methyl, which was, owing to its thiolo structure, also a more potent inhibitor of cholinesterase than the other two pesticides applied.

Whole blood and plasma enzyme activities were found to be unrelated to the urinary excretion of OP pesticide metabolites. The only exception was one mixer after work with Demeton-S-methyl with 26 and 42% of cholinesterase activity in whole blood and plasma respectively, and at the same time with the highest concentration of DMPThK in urine.

Mixers, as the most exposed group, judging from urinary excretion of metabolites after work with Azinphos-methyl or Methidathion, did not have significantly different cholinesterase activity compared to applicators and mechanics or field workers. Even the highest metabolite concentrations in individual mixers were not accom-

TABLE IV

Cholinesterase activity in whole blood and plasma in mixers (I), applicators (II) and other workers in the orchard (III) at the end of the third day of work with Demeton-S-methyl.

| | Cholinesterase activity (% pre-exposure value) | | | | | |
| | Whole blood | | | Plasma | | |
| | I | II | III | I | II | III |
|---|---|---|---|---|---|---|
| Number of workers | 7 | 6 | 7 | 6 | 5 | 6 |
| Median | 97 | 118 | 97 | 93 | 112 | 101 |
| Range | 26–126 | 94–129 | 86–128 | 42–116 | 95–118 | 96–125 |
| Mean $\pm$ S.D. | $91 \pm 32$ | $116 \pm 17$ | $99 \pm 15$ | $90 \pm 26$ | $109 \pm 10$ | $104 \pm 10$ |
| F-ratio (Analysis of variance) | | $1.15^a$ | | | $5.48^b$ | |

[a] Not significantly different at the 5% level.
[b] Significantly different at the 5% level.

TABLE V

Cholinesterase activity in whole blood and plasma in mixers (I), applicators (II) and mechanics (III) one day after the end of work with Azinphos-methyl.

| | Cholinesterase activity (% pre-exposure value) | | | | | |
| | Whole blood | | | Plasma | | |
| | I | II | III | I | II | III |
|---|---|---|---|---|---|---|
| Number of workers | 6 | 10 | 5 | 6 | 10 | 5 |
| Median | 98 | 95 | 103 | 89 | 96 | 93 |
| Range | 82–106 | 81–110 | 64–114 | 63–91 | 78–120 | 84–97 |
| Mean $\pm$ S.D. | $95 \pm 10$ | $96 \pm 9$ | $95 \pm 20$ | $82 \pm 13$ | $94 \pm 13$ | $92 \pm 5$ |
| F-ratio (Analysis of variance) | | $0.35^a$ | | | $2.78^a$ | |

[a] Not significantly different at the 5% level.

TABLE VI

Cholinesterase activity in whole blood and plasma in mixers (I), applicators (II) and field workers (III) one day after the end of work with Methidathion.

| | Cholinesterase activity (% pre-exposure value) | | | | | |
| | Whole blood | | | Plasma | | |
| | I | II | III | I | II | III |
| --- | --- | --- | --- | --- | --- | --- |
| Number of workers | 6 | 12 | 8 | 6 | 10 | 7 |
| Median | 83 | 86 | 82 | 95 | 101 | 91 |
| Range | 67–96 | 65–108 | 62–102 | 75–101 | 81–115 | 74–114 |
| Mean ± S.D. | 84±11 | 86±13 | 83±13 | 92±10 | 99±10 | 92±12 |
| F-ratio (Analysis of variance) | | 0.27[a] | | | 0.13[a] | |

[a]Not significantly different at the 0.5% level.

panied by greatly depressed cholinesterase activity. In both the mixer with the highest and the mixer with the lowest urinary concentration of Azinphos-methyl metabolites enzyme inhibition, only in the plasma, was nearly identical i.e. 31 and 37%.

On the other hand, in nine out of 20 workers exposed to Demeton-S-methyl the whole blood (two mixers, five applicators and two other workers) and plasma (two mixers, four applicators and three other workers) cholinesterase activity was higher than 100%. Similarly, an activity above the baseline value was determined in nine out of 22 workers exposed to Azinphos-methyl (two mixers, four applicators and three mechanics) in whole blood and in three workers (all applicators) in plasma. After exposure to Methidathion in four out of 26 workers (three applicators and one field worker) and in seven out of 23 workers (one mixer, five applicators and one field worker) the enzyme activity was higher than 100% both in whole blood and plasma.

Obviously, variations in cholinesterase activity in whole blood and plasma could hardly be related only to the absorption of one particular pesticide, partially because of successive multiple exposures of workers to different OP pesticides during the spraying season but also because of additional impacts which were not known for each worker.

## CONCLUSION

Urinary excretion of the metabolites of Demeton-S-methyl, Azinphos-methyl and Methidathion in different groups of agricultural workers was confirmed as a reliable and sensitive indicator of subtle differences in exposure during the application of these pesticides in an apple-orchard (Figure 1a). Of the two determined specific metabolites of Azinphos-methyl and Methidathion DMTPK was found to be present in most urine samples and in considerably higher concentrations than DMDTPK. Similar results were reported for formulators exposed to a variety of OP pesticides most of whom had dialkyl phosphorothionates present in their urines in concentrations higher than the corresponding phosphorodithioates.[13,14]

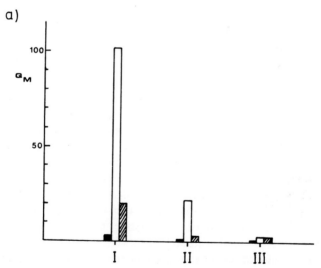

FIGURE 1(a)   The ratio of post- to pre-exposure median concentrations ($Q_M$) of the urinary metabolite of Demeton-S-methyl (■ DMPThK), Azinphos-methyl (□ DMTPK) and Methidathion (▨ DMTPK) in mixers (I), applicators (II) and other workers (III).

b)

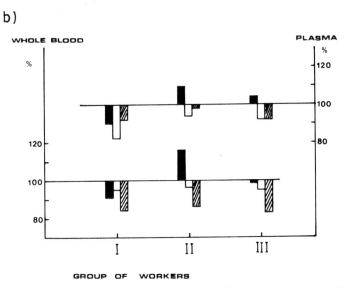

FIGURE 1(b)   Mean decrease/increase (%) in baseline cholinesterase activity in whole blood and plasma in mixers (I), applicators (II) and other workers (III) after exposure to Demeton-S-methyl (■), Azinphos-methyl (□), or Methidathion (▨).

It seems that the metabolic degradation of absorbed dithiophosphoric acid triesters results preferably in an increased urinary excretion of phosphorothionate which is consequently a more sensitive but less specific measure of exposure than phosphorodithioate.

Different absorption of OP pesticides in several groups of workers was not, as the rule, reflected by pertinent differences in their plasma and whole blood cholinesterase activities (Figure 1b). The amounts absorbed were not sufficient to cause a major reduction of enzyme activity in exposed workers except in one mixer exposed to Demeton-S-methyl. The possibility of an accelerated rate of enzyme reactivation in the body due to successive multiple exposures of workers to small amounts of OP pesticides cannot be disregarded.[16]

Monitoring of specific urinary metabolites contribute to the early detection of defectivenesses in protection measures at special workplaces. The workers showing excessive urinary excretion of one or more pesticide metabolites should be better protected or removed from further exposure before a major inhibition of cholinesterase activity or other symptoms of poisoning appear.

## Acknowledgements

This project was supported by the Environmental Protection Agency (U.S.A.) through funds made available to the U.S. Yugoslav Joint Board on Scientific and Technological Cooperation.

The authors gratefully acknowledge the skilful technical assistance of Mrs. M. Kramarić and M. Matašin in analysis of cholinesterase activity in blood samples. Special thanks are due to Professor Zlata Štefanac for valuable discussions.

## References

1. A. L. Hayes, R. A. Wise and F. W. Weir, *Am. Ind. Hyg. Assoc. J.* **41**, 568 (1980).
2. J. B. Knaak, K. T. Maddy, T. Jackson, A. S. Frederickson, A. S. Peoples and R. Love, *Toxicol. Appl. Pharmacol.* **45**, 755 (1978).
3. M. A. Quinones, J. D. Bogden, D. B. Louria, A. El Nakah and C. Hansen, *Sci. Total Environ.* **6**, 155 (1976).
4. B. Svetličič and K. Wilhelm, *Arh. hig. rada* **24**, 357 (1973).
5. E. Fournier, M. Sonnier and S. Dally, *Clin. Toxicol.* **12**, 457 (1978).
6. A. F. Machin, M. P. Quick and D. F. Waddel, *Analyst* **98**, 176 (1973).
7. D. Blair and H. R. Roderick, *J. Agric. Food Chem.* **24**, 1221 (1976).
8. D. E. Bradway and T. M. Shafik, *J. Agric. Food Chem.* **25**, 1342 (1977).
9. J. E. Davies, H. F. Enos, A. Barquet, C. Morgade and J. X. Danauskas. In: *Toxicology and Occupational Medicine* (W. B. Deichman, ed.) (Elsevier/North-Holland, New York, 1979), Vol. 4, pp. 369–378.
10. V. Drevenkar, Z. Fröbe, Ž. Vasilić, B. Tkalčević and Z. Štefanac, *Sci. Total Environ.* **13**, 235 (1979).
11. V. Drevenkar, B. Štengl, B. Tkalčević and Ž. Vasilić, *Intern. J. Environ. Anal. Chem.* **14**, 215 (1983).
12. J. B. Knaak, K. T. Maddy and S. Khalifa, *Bull. Environm. Contam. Toxicol.* **21**, 375 (1979).
13. T. Shafik, E. D. Bradway, H. F. Enos and A. R. Yobs, *J. Agric. Food Chem.* **21**, 625 (1973).
14. D. Y. Takade, J. M. Reynolds and J. H. Nelson, *J. Agric. Food Chem.* **27**, 746 (1979).
15. M. Zietek, *Mikrochim. Acta (Wien)* **II**, 75 (1979).
16. S. A. Peoples and J. B. Knaak, *ACS Symp. Ser. (Pesticide Residues Exposure)* **182**, 41 (1982).
17. D. P. Morgan, H. L. Hetzler, E. F. Slach and I. L. Lin, *Arch. Environm. Contam. Toxicol.* **6**, 159 (1977).
18. A. I. Vogel, *Practical Organic Chemistry* (Longmans, Green and Co., London, New York, Toronto, 1956), p. 971.
19. G. L. Ellman, K. D. Courtney, A. Valentino Jr. and R. M. Featherstone, *Biochem. Pharmacol.* **7**, 88 (1961).
20. C. G. Daughton, D. G. Crosby, R. L. Garnas and P. H. Hsieh, *J. Agric. Food Chem.* **24**, 236 (1976).
21. B. Petz, *Osnovne statističke metode za nematematičare* (Sveučilišna naklada Liber, Zagreb, 1981), p. 321.

# A Test System for the Determining of the Fate of Pesticides in Surface Water†

## Protocol and Comparison of the Performance for Parathion of Ecocores and Micro Ecosystems from Two Sources

N. W. H. HOUX and A. DEKKER

Institute for Pesticide Research, P.O. Box 650, 6700 AR Wageningen, The Netherlands

(Received July 9, 1986; in final form August 20, 1986)

Protocols are described for constructing laboratory micro-ecosystems (MES), incubating them with radiolabelled pesticides and then using a routine test procedure to ascertain the fate of these pesticides in surface water. The performance of fresh ecocores and acclimated MES from two sources was compared. The influence of the duration of the acclimation to room temperature and a light cycle on the fate of parathion was studied. The variation between replicates of MES was less than that between ecocores. The eutrophic ecocores and MES performed similarly, the oligotrophic ecocores transformed parathion faster than the oligotrophic MES. In eutrophic systems, reduction to aminoparathion was much faster than in oligotrophic systems. The sandy oligotrophic MES needed a longer acclimation to laboratory conditions than the eutrophic MES to produce reproducible results. The results of year-to-year experiments were also more reproducible for the eutrophic MES.

KEY WORDS: Test procedure, fate, parathion, aquatic micro ecosystem.

†Presented at the Workshop on Chemistry and Fate of Organophosphorus Compounds, Amsterdam, Holland, June 18–20, 1986.

This article was first published in International Journal of Environmental Analytical Chemistry, Volume 29, Numbers 1+2 (1987).

## INTRODUCTION

Large areas of economically important Dutch agricultural and horticultural land are traversed by ditches and canals of sluggish water. This hydrological system is necessary to cope with the surplus of rainfall and because much of the country is below sealevel. Not surprisingly, therefore, when a pesticide is submitted for registration in the Netherlands, data on its fate in surface water are mandatory. "Fate" is defined as the process of transport, transformation and final disposal by natural means.

For an ecotoxicological evaluation of the environmental loading by pesticides, the conditions for test systems in the laboratory should resemble those of the complex natural systems as closely as possible. This means that at least the naturally present degraders (bacteria, fungi, algae, etc.) should be allowed to develop into a micro-ecosystem in an undisturbed microcosm of sediment and water under aeration of the surface layer and in a day/night light cycle.

J. M. Giddings[1] elaborated a protocol for using aquatic pond microcosms for determining the fate of micropollutants in eco-systems and for studying some of their effects. In his excellent study he concluded that after several weeks of equilibrium under constant laboratory conditions, 70 l microcosms of sediment, water and a complete biotic community can develop into useful test systems that are a realistic simulation of the parent ecosystem. Smaller (7 l) microcosms behaved similarly but were less stable and more easily disturbed.

Routine work requires many replicates of small systems that can be analysed easily and economically. An equilibrated laboratory micro-ecosystem cannot be subsampled without disturbing the complex ecosystem that depends on numerous niches in gradients of light, nutrients, temperature, pH, oxygen and redox potential.

The objective of our study was to develop a routine test system comprising an aquatic micro-ecosystem containing the essential functional components of the parent natural ecosystem, and to develop subsequent analytical procedures to provide data on the fate of pesticides in surface water.

Because of its short half life, the abundance of information on its degradation in aqueous systems,[2] and its easy availability, $^{14}C$-parathion was used as a substance to test the model. Using the fate of parathion and its transformation products as criterion, attention was paid to the following aspects:

1) Constructing replicate laboratory micro-ecosystems (MES);

2) Comparing fresh ecocores with acclimated MES;

3) Ascertaining the influence of the age of the MES on performance;

4) Developing adaptable and economical analytical procedures.


## MATERIAL AND METHODS

### Constructing the micro-ecosystem

Sediment and water were collected from a ditch in an apple orchard and a small man-made lake near Wageningen. About 25 l of the uppermost 5 cm of the sediment was scooped into a bucket, and the same day was passed through a sieve (2 mm mesh) in the laboratory, with the corresponding surface water. The sieved sediment was allowed to settle overnight in an open plastic box (60 × 30 × 20 cm). The water was carefully drained off and after mixing, sediment samples were taken so that the physical and chemical properties of the sediment could be determined (Table I).

At the start of our study the sieved sediment was intensely emulsified with an Ultra Turrax homogenizer (Janke & Kunkel, type T 45 S7) to improve the replicability of the weighing of the wet sediment and the sampling of the extracted sediment for combustion afterwards. The experiment on the acclimation of the MES before the incubation was performed with emulsified sediment.

Portions of wet sediment (from 5 to 10 g dry weight) taken with a glass corer were weighed into individually numbered and tared glass centrifuge tubes (70 ml) equipped with a screw thread. The tubes were filled with the drained water, taking care not to disperse the sediment layer, and were then randomly placed in a glass rack in a glass aquarium (80 × 40 × 20 cm) with a layer (3 cm thick) of the sieved sediment on the bottom and with the corresponding surface water standing 2–3 cm over the open tubes. The aquarium was covered with a glass lid (Figure 1).

## TABLE I

Physical and chemical properties of the sediments studied.[a]

| Site | pH-kCl | Organic matter (%) | CaCO$_3$ (%) | Clay <2 μm (%) | Silt 2–50 μm (%) | Sand >50 μm (%) | Total P (%) | Total N (%) | Dry[b] weight (105°C) (%) | Fresh[b] weight of sample (g) |
|------|--------|--------------------|--------------|----------------|------------------|-----------------|-------------|-------------|---------------------------|-------------------------------|
| Ditch | 7.2 | 6.4 | 3.9 | 23 | 40 | 27 | 0.56 | 0.37 | 34.2 | 19.5 |
| Lake | 7.6 | 0.8 | 3.9 | 6 | 10 | 79 | 0.08 | 0.05 | 73.3 | 13.5 |

[a]Determined by the Laboratory for Soil and Crop Testing, Oosterbeek, The Netherlands.
[b]Determined by the Institute for Pesticide Research, Wageningen, The Netherlands.

## MICRO ECOSYSTEM ASSEMBLY

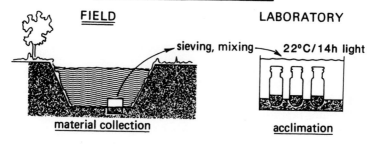

## INCUBATION

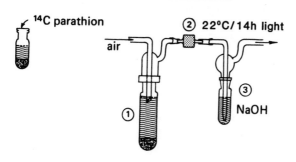

1 micro ecosystem
2 sep-pak $C_{18}$ : volatiles trap
3 0,5N NaOH : $^{14}CO_2$ trap

FIGURE 1   Set-up for acclimation and incubation of the micro-ecosystem.

D

Under identical conditions at room temperature and 14 hrs cool white fluorescent light per day, 108 tubes with a known amount of sediment (variation 0.5 g dry weight) were allowed to develop into replicate MES as part of a larger ecosystem of 3 cm of sediment overlain by 16 cm of water with a total volume of 60 l. Periodically, excess water plants, algae or snails were removed and the evaporated water was topped up.

## Ecocores

In the field open glass tubes (20 cm long, 3 cm inner diameter) were pressed 6 cm deep into the sediment that was overlain by shallow surface water (approx. 30 cm deep), next to each other. As the tubes were being pushed into the sediment, the sediment cores in the tubes usually sank about 2 cm and sometimes bubbles of gas escaped. The upper end of each tube was closed with a rubber stopper and the tube was carefully pulled up. While still under water, the lower end of the tube was also plugged with a rubber stopper.

In the laboratory the water was drained off to a level 3 cm above the surface of the sediment. The incubation and further handling were the same as described for the MES. The ecocores were incubated on the day of collection and five replicates were used for each estimation.

## Incubation

At appropriate times the MES tubes were taken from the aquariums and their outer surface was cleaned. All handling was done very carefully to minimize the disturbance of the MES. Water was drawn off until the sediment and the water weighed just under 40 g. The radiolabelled test substance was applied in a dose of 10 ml of distilled water so that the final concentration in the MES matched the concentration expected in field situations (in our study, from 1 to 2 mg parathion per litre). The total weight of the MES was made up to 50 g with water.

The dosing solution was made by putting calculated amounts of hexane solutions of unlabelled parathion (0,0-diethyl 0-4-nitrophenyl phosphorothioate, water solubility $24 \, \text{mg} \, \text{l}^{-1}$) and [ring-2,6-$^{14}$C] parathion (Amersham, spec. act. $2.67 \, \text{M} \, \text{Bq} \, \text{mg}^{-1}$) in a glass bottle.

After the hexane had evaporated, an appropriate volume of distilled water was added and the glass stoppered bottle was shaken for at least 2 hours or sonicated (Bransonic, Model 52, 50,000 Hz) for 15 minutes. If the radioactivity measurement indicated that the parathion was not yet fully dissolved (in our study ca $10^8$ dpm and between 5 and $10 \, \text{mg} \, l^{-1}$) the treatment was continued. Finally, the concentration was determined by HPLC with UV and radiometric detection.

During the incubation each MES with its traps for volatiles and $CO_2$ was mounted on an aluminium stand. The air perfusion assembly was screwed air-tight on the centrifuge tube with a silicon-rubber ring. Air was bubbled through the 5 mm top layer of the MES water at a rate of $1 \, \text{ml} \, \text{min}^{-1}$ and subsequently through the volatile trap (SEP PAK$^{R}$, Waters) and the $CO_2$ trap (5 ml of 0.5 N NaOH) (Figure 1). At least two replicates were performed for each estimation during incubation at room temperature and 14 hrs light per day.

## Analysis

At 2, 7 and 14 days of incubation, replicate systems were analysed. The $^{14}C$ activity from aliquots of all fractions was measured by liquid scintillation counting in Optifluor (United Technologies Packard) with a Liquid Scintillation Analyser (Phillips Model PW 4540). As no radioactivity from $^{14}C$-parathion was detected in the methanol eluate from the SEP PAK$^{R}$ cartridges and the eluted $C_{18}$ adsorbent, these volatile traps were omitted in later experiments with parathion.

The amount of $^{14}CO_2$ that had formed was measured in 0.5 ml of the NaOH trap. The MES tube was centrifuged (10 min, 2,200 rpm, Heraeus Christ Varifuge GL), the weight recorded and the radio-activity in 0.5 ml of the water assessed (Figure 2). The water layer was extracted by suction through a 3 ml Baker SPE $C_{18}$ disposable column.

A special vacuum box was constructed for handling up to 12 extraction columns simultaneously during the prewash, extraction and eluation procedures. This device and the exact procedures will be described elsewhere.

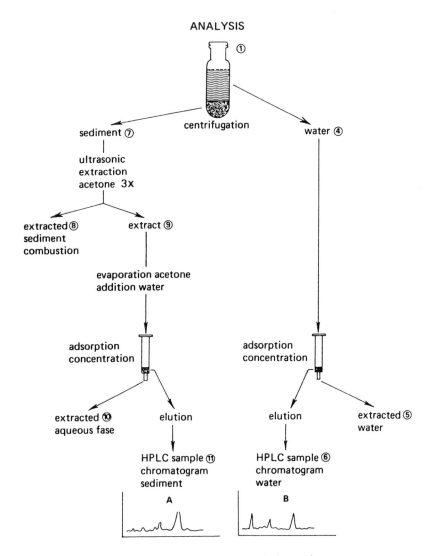

FIGURE 2   Outline of the analytical procedure.

The volume of the eluate was recorded and a 0.5 ml aliquot taken for radioassay. The adsorbent of the column was extracted with 3 ml of methanol directly in the HPLC sample vial. After adding of 1 ml of distilled water and mixing, the radioactivity of a 100 μl aliquot was measured. The weight of the wet sediment was recorded, 25 ml of acetone added, the tube closed with a screw cap and vigorously shaken by hand for a few seconds to disperse the sediment. The sediments were extracted by sonication during 10 min. The extract was collected in a graduated cylinder. This extraction procedure was repeated twice with 25 ml of 10% water in acetone. The volume of the combined extracts was recorded and the radioactivity in an aliquot was measured. The acetone was evaporated in a vacuum rotator until the first drops of water appeared in the condensor. The solids were washed from the wall of the flask with 1 to 2 ml of acetone and 40 ml of distilled water was added. The yellow to dark brown and turbid aqueous phase was extracted by the same extraction column procedure as described for the MES water, and the radioactivity in the eluate and in the HPLC sample was measured.

Only a few per cent of the radioactivity present in the aqueous eluates from the extraction columns proved to be extractable by subsequent extraction with chloroform or ethylacetate. So this rapid extraction, concentration and partial clean-up procedure extracted all non-polar components.

The extracted sediment was dried overnight at 105°C, homogenized in a mechanical mortar (Fritsch, Pulverisette 2) and the $^{14}CO_2$ radioactivity determined after two aliquots (approx. 200 mg each) had been combusted in a Packard Model 306 Sample Oxidizer.

The values found for the radioactivity of the volatile trap, the $^{14}CO_2$ trap, the MES water, the extracted sediment and the sediment extract were used to calculate the total $^{14}C$ recovery.

The HPLC apparatus was equipped with a Waters Model E 590 solvent delivery system, a Model 710 A WISP sample processor, a column oven, a Schoeffel model 770 spectrophotometric detector (used at 283 nm), a Berthold LB 5026 radioactivity monitor with a HS solid scintillator detection cell and with an Apple IIe computer with an Isomess IM 2013 Tripletrace Radio Chromatographic Program. Automatically 250 μl samples were injected into a 50 × 4.6 mm guard column (Copell ODS, Whatman) and a

$150 \times 4.6$ mm analytical column (Lichrosorb 5 $RP_{18}$, Chrompack) at 50°C in phosphate buffer/methanol/acetonitrile $(30 + 65 + 5)$ at a flow rate of $1 \, \text{ml} \, \text{min}^{-1}$. The phosphate buffer (pH 9) was 0.01 M in the mobile phase, methanol and acetonitrile were "Baker Analyzed" HPLC reagents. Typical retention times under these conditions were parathion 7.8 min, aminoparathion 4.7 min, *p*-aminophenol 3.4 min and *p*-nitrophenol 3.0 min.

## RESULTS

Data from a single experiment with fresh ecocores, incubated from the day of their collection, were compared with those from two experiments with MES constituted from material collected from the same sites and on the same days, but acclimated to laboratory conditions for 35 or 41 days and 63 or 69 days (Tables II and III). Small invertebrates, e.g. Cyclops spp. Tubifex spp. and leeches were active in the ecocores during the incubation periods studied. The total $^{14}C$ recoveries of these experiments were satisfactory and only a slight mineralization was found in all systems. The replicability of the MES was much better than that of the ecocores; this is clearly shown by the larger variation in the measurements (Figure 3) and the standard deviations (Tables II and III).

During the incubation the radioactivity in the water fell and that in the sediment rose. This latter increase was mainly due to the increase in the non-extractable fraction.

The non extractable radioactivity in the lake ecocores was much greater than that in the lake MES. The opposite tended to be found for the ditch systems (Tables III and IV). An almost constant fraction of the radioactivity was extractable from the ditch systems during the incubation. In the older lake MES, however, this fraction was higher than in the younger MES and ecocores.

The persistence of parathion and transformation products in the extracts from water and sediments is given in Tables IV and V and Figure 4.

The polar products had retention times of between 2 and 3.5 minutes in the HPLC chromatograms, i.e. less than aminoparathion. The intermediate products were situated between the amino-parathion and parathion peaks. No identification was undertaken

TABLE II

Recovery of radioactivity in percentage of dose of $^{14}C$-parathion, during incubation of ditch systems.[a]

| System acclimation time | Incubation time (days) | $^{14}CO_2$ | Water | Sediment | | Ecotube rinsing | Total |
|---|---|---|---|---|---|---|---|
| | | | | Extractable | Non-extractable | | |
| Ecocore 0 days | 2 | nc[b] | 65.4 ± 3.7 | 23.5 ± 2.5 | 11.7 ± 1.7 | 0.7 ± 0.2 | 101.3 ± 2.6 |
| | 7 | nc | 33.3 ± 11.7 | 23.7 ± 3.1 | 37.6 ± 5.5 | 0.5 ± 0.1 | 95.2 ± 4.5 |
| | 14 | 1.6 ± 1.6 | 28.2 ± 7.8 | 19.1 ± 3.8 | 40.8 ± 10.5 | 0.5 ± 0.2 | 90.2 ± 1.2 |
| MES 41 days | 2 | <0.05 | 74.0 ± 1.1 | 18.9 ± 0.9 | 8.3 ± 2.2 | | 101.2 ± 0.4 |
| | 7 | 0.6 ± 0.1 | 45.6 ± 4.5 | 25.0 ± 1.0 | 24.6 ± 3.9 | | 95.7 ± 0.4 |
| | 14 | 1.0 ± 0.6 | 22.1 ± 2.7 | 24.4 ± 1.5 | 45.9 ± 2.8 | | 93.4 ± 0.8 |
| MES 69 days | 2 | <0.05 | 67.8 ± 1.1 | 20.8 ± 1.8 | 9.9 ± 0.9 | | 98.5 ± 1.8 |
| | 7 | 0.4 ± 0.1 | 36.0 ± 2.0 | 26.7 ± 2.1 | 35.4 ± 4.7 | | 98.6 ± 0.9 |
| | 14 | 0.6 ± 0.3 | 18.0 ± 0.4 | 22.9 ± 0.8 | 53.9 ± 1.5 | | 95.3 ± 0.4 |

[a] $\bar{x} \pm s$, ecocores $n=5$, MES $n=3$.
[b] Not collected.

## TABLE III

Recovery of radioactivity in percentage of dose of $^{14}C$ parathion, during incubation of lake systems.[a]

| System acclimation time | Incubation time (days) | $^{14}CO_2$ | Water | Sediment Extractable | Non-extractable | Ecotube rinsing | Total |
|---|---|---|---|---|---|---|---|
| Ecocore 0 days | 2 | nc[b] | 76.9 ± 7.0 | 12.4 ± 3.9 | 8.9 ± 2.8 | 0.5 ± 0.0 | 98.7 ± 0.7 |
| | 7 | nc | 44.9 ± 2.2 | 14.7 ± 0.9 | 31.6 ± 2.0 | 0.7 ± 0.1 | 91.9 ± 0.7 |
| | 14 | 1.0 ± 1.0 | 23.5 ± 6.5 | 14.2 ± 2.7 | 49.1 ± 4.3 | 0.5 ± 0.2 | 88.2 ± 3.6 |
| MES 35 days | 2 | <0.05 | 76.7 ± 2.5 | 16.3 ± 0.3 | 3.4 ± 0.4 | | 86.0 ± 2.0 |
| | 7 | 0.4 ± 0.3 | 57.8 ± 1.1 | 19.7 ± 0.6 | 15.2 ± 0.2 | | 93.1 ± 1.7 |
| | 14 | 3.3 ± 0.5 | 42.1 ± 0.8 | 15.6 ± 0.4 | 31.9 ± 1.1 | | 93.0 ± 1.4 |
| MES 63 days | 2 | <0.05 | 70.9 ± 2.3 | 22.2 ± 1.4 | 3.0 ± 0.2 | | 96.1 ± 0.8 |
| | 7 | 0.5 ± 0.2 | 55.0 ± 0.7 | 25.7 ± 0.7 | 13.7 ± 0.6 | | 94.9 ± 1.3 |
| | 14 | 3.5 ± 0.1 | 42.6 ± 3.6 | 21.4 ± 1.9 | 26.4 ± 0.9 | | 93.9 ± 0.9 |

[a] $\bar{x} \pm s$, ecocores $n = 5$, MES $n = 3$.
[b] Not collected.

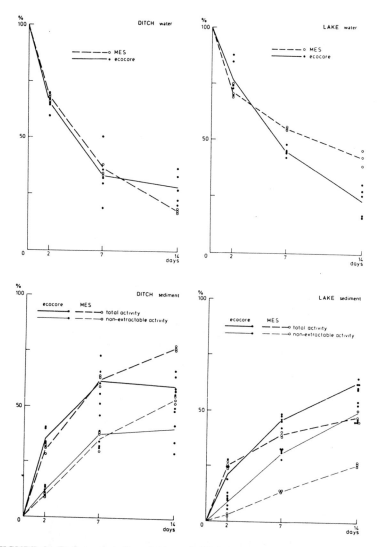

FIGURE 3 Performance of ecocores and MES from ditch and lake. Remaining total radioactivity in water and sediments and non-extractable radioactivity in sediments after 2, 7 and 14 days of incubation.

TABLE IV

Parathion and transformation products in ditch systems in percentage of dose.[a]

| System acclimation time | Incubation time (days) | Water extract | | | Sediment extract | | |
|---|---|---|---|---|---|---|---|
| | | Polar products | Amino-parathion | Parathion | Amino-parathion | Intermediate products | Parathion |
| Ecocore 0 days | 2 | 3.7±0.8 | 13.2±3.3 | 29.0±6.5 | 7.3±1.4 | 2.3±0.7 | 9.2±2.9 |
| | 7 | 3.4±0.8 | 18.0±5.1 | 5.7±2.7 | 8.9±1.8 | 2.8±1.5 | 3.5±0.9 |
| | 14 | 4.3±0.6 | 17.3±6.0 | —[b] | 11.5±2.4 | — | 1.4±0.5 |
| MES 41 days | 2 | 2.5±0.6 | 31.7±8.3 | 30.9±11.4 | 5.8±0.8 | — | 3.1±0.8 |
| | 7 | 6.4±0.7 | 29.7±2.6 | 2.2±0.4 | 8.7±0.6 | — | 1.8[c] |
| | 14 | 4.9±0.6 | 12.8±1.8 | — | 12.5±0.9 | — | — |
| MES 69 days | 2 | 4.6±0.3 | 30.9±2.9 | 26.6±4.3 | 6.9±1.2 | — | 5.6±0.4 |
| | 7 | 4.3±0.1 | 29.7±1.3 | — | 14.4±0.7 | 2.3±0.9 | 0.9±0.5 |
| | 14 | 4.9±0.2 | 10.6±0.7 | — | 11.9±0.7 | — | — |

[a] $\bar{x} \pm s$, ecocores $n = 5$, MES $n = 3$.
[b] Not detected.
[c] Single measurement.

TABLE V

Parathion and transformation products in lake systems in percentage of dose.[a]

| System acclimation time | Incubation time (days) | Water extract | | | Sediment extract | | |
|---|---|---|---|---|---|---|---|
| | | Polar products | Amino-parathion | Parathion | Amino-parathion | Intermediate products | Parathion |
| Ecocore 0 days | 2 | 8.8±0.8 | 7.5±2.0 | 55.9±8.3 | 2.5±0.9 | —[b] | 6.0±2.6 |
| | 7 | 7.6±1.6 | 20.8±1.5 | 11.6±1.6 | 5.9±1.3 | — | 3.4±1.0 |
| | 14 | 5.9±1.5 | 13.8±6.0 | — | 8.2±1.7 | — | — |
| MES 35 days | 2 | — | — | 67.5±1.8 | — | — | 15.5±0.3 |
| | 7 | 2.4±1.1 | 9.0±0.5 | 42.3±2.7 | 5.2±0.4 | — | 12.7±0.5 |
| | 14 | 2.4±0.6 | 13.2±1.8 | 21.6±1.7 | 3.6±0.9 | — | 9.2±0.3 |
| MES 63 days | 2 | 2.3±0.1 | 2.5[c] | 65.4±4.0 | — | — | 20.7±1.6 |
| | 7 | 2.9±0.7 | 8.0±1.8 | 41.0±4.2 | 2.7±0.3 | — | 15.7±3.6 |
| | 14 | 3.0±0.7 | 11.1±0.2 | 23.8±2.9 | 3.8±0.7 | — | 12.5±1.5 |

[a] $\bar{x} \pm s$, ecocores $n = 5$, MES $n = 3$.
[b] Not detected.
[c] Single measurement.

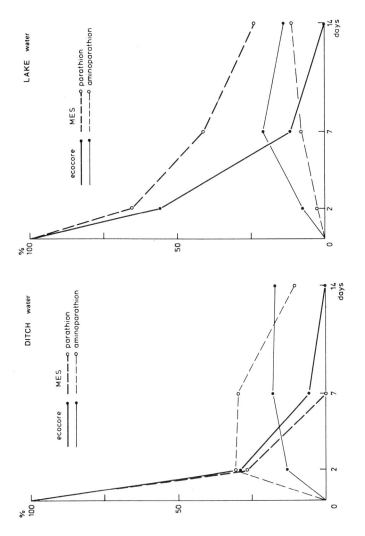

FIGURE 4 Performance of ecocores and MES from ditch and lake. Persistence of parathion and aminoparathion in water after 2, 7 and 14 days of incubation.

and even the aminoparathion peak was not homogeneous and contained other radiolabelled products.

Between 3.7% and 4.7% of the radioactivity remained in the extracted water from the ditch ecocores and between 1.4% and 2.9% in that from the ditch MES. The corresponding fractions of the lake systems contained about 1% more radioactivity. Thus, only a minor fraction of the transformation products consisted of very polar components.

It should be mentioned here that for unknown reasons about 20% of the radioactivity was lost during the extraction of the first ecocore water sample processed, notably the ditch ecocores after 2 days of incubation.

Parathion disappeared rapidly from the water in the ditch systems concomitantly with its transformation into aminoparathion. This latter component decreased gradually in the water of the MES.

Almost twice the amount of parathion was present in the ditch ecocore sediment than in the MES sediment (Table IV). This may have been caused by the turbulent activity of Tubifex worms that were present only in the loose organic upper layer of the ditch ecocore sediments. The percentage of aminoparathion increased gradually in the sediments of the younger ditch MES and ecocores, but reached a maximum of 14.4% at 7 days in the older MES.

A totally different pattern was found in the lake systems (Table V, Figure 4). After 2 days, about 60% of the radioactivity in the water was still present as parathion, and more than 20% remained in the MES water after 14 days. The aminoparathion fraction had peaked in the lake ecocore water, but was in an early stage of build up in the lake MES water. The polar products formed a remarkably high fraction in the lake ecocore water. In the lake MES sediment the percentage of the parathion was the highest of all systems, whereas that of aminoparathion was the lowest.

Earlier experiments had indicated that the MES performance changed gradually during the acclimation period. In our study trends could be seen by comparing ditch MES acclimated for 41 and 69 days (Tables II and IV), the clearest being the percentage of [14]C in water, the non-extractable fraction of sediment and the amount of aminoparathion in the sediment. In the present study we therefore investigated the influence of ageing of the MES on its performance with the fate of parathion as criterion. Sieved and emulsified sediments ($10 \pm 0.3$ g dry weight) and surface water (40 g) collected from both sites were acclimated to laboratory conditions. Incubations were performed at 1, 3, 5, 8, 16, 26 and 71 weeks. In Figure 5 the percentages of total $^{14}C$, parathion and aminoparathion show trends until about 26 weeks of acclimation in the case of the lake MES and 16 weeks in the case of the ditch MES water.

Tʰe aminoparathion curves during the ageing of the lake and ditch MES are compared in Figure 6. The lake MES mainly showed a build up of the aminoparathion content, with only a small decline between 7 and 14 days. The ditch MES showed an increasing amount of aminoparathion at 2 days of incubation and an increasing rate of decline from 2 days onwards.


## DISCUSSION

The rapid reduction of parathion to aminoparathion in the water and the sediment and the rapid increase of the non-extractable fraction in the sediment agree with the well documented transformation of parathion by algae[3] and anaerobic micro-organisms[2,4,5] and the strong adsorption of aminoparathion in soils.[6] Acclimation and incubation in a light regime are necessary to provide the natural elements of ecosystems (the production of oxygen and organic nutrients in the water columns, the degrading capacity of algae and their associated biomass, and the diurnal changes in pH). In our ditch aquariums the largest fluctuation in pH (from 7 to 9) was measured after several weeks of acclimation. Within 4 to 7 days an $E_h$ below $-50$ mV was measured in the sediments. To support the autotrophic organisms, perfusion with compressed air containing a normal $CO_2$ content was preferred above the use of $CO_2$-free air during incubation.

In our search for an economical test procedure we attempted to miniaturize MES by acclimating replicates of 70 ml microcosms in a 70 l MES. The use of centrifuge tubes for the determination of the persistence of pesticides in an estuarine environment has recently been described.[7] However, that research was not aimed at an ecosystem test, and unacclimated sediment and seawater were agitated continuously at 100 rpm on a rotator table. The individual acclimation of MES-tubes to laboratory conditions has also been

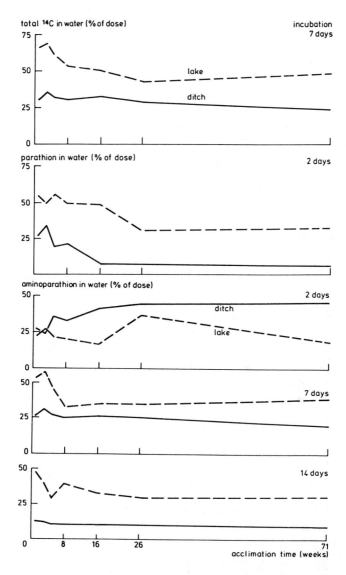

FIGURE 5  Effect of acclimation on performance of homogenized lake and ditch MES. Incubation after 1, 3 5, 8, 16, 26 and 71 weeks of acclimation to laboratory conditions. Contents in water of total $^{14}$C after 7 days, parathion after 2 days and aminoparathion after 2, 7 and 14 days of incubation.

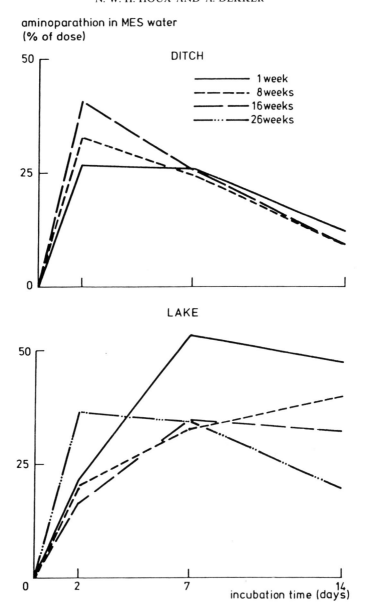

FIGURE 6   Persistence patterns of aminoparathion in water of lake and ditch MES
after acclimation to laboratory conditions for 1, 8, 16 and 26 weeks.

described.[8] However, in that study, glass ecocore tubes, closed at both ends with rubber stoppers and thus not part of a larger ecosystem, were used. No advantage for the subsequent chemical analysis was obtained and the sediment had to be transferred for extraction.

In our experiments the replicability of all acclimated MES of the same batch was very good and better than that of the fresh ecocores. In our set-up a longer acclimation to laboratory conditions (16 to 26 weeks) is indicated than in one non-compartmentalized 70 l aquarium, where at least 6 weeks of acclimation was necessary.[2] Our data conform with the longer equilibrium time found for aquariums of 7 l.[2] Unlike the lake MES, our ditch MES performed similarly or even better than the fresh ecocores. The ecocores were collected early September 1985 when outdoor temperature was approximately 18°C and there was 13 hours daylight.

The reproducibility of the system's performance was examined by comparing the aminoparathion content in the water of homogenized MES prepared in October 1984 (outdoor temperature approx. 14°C, 11 hrs daylight) and acclimated for 8 weeks, with that in the water of 63 and 69 days acclimated 1985 MES. An excellent reproducibility was obtained for the ditch MES: the aminoparathion contents for 2, 7 and 14 days of incubation were 33, 25 and 9% in 1984 and 33, 27 and 12% in 1985. The corresponding data for the lake systems were 20, 33 and 40% in 1985 versus 1, 19 and 18% in 1985.

Our data agree with the conclusion reached by Giddings[2] that a microcosm with sand as sediment does not develope into an MES as readily as one with a richer pond sediment, and that sandy MES are less stable. It is therefore doubtful if sandy sediments from oligo-trophic waters can be used for this test. Even if these poor sandy MES simulate the performance in an oligotrophic surface water, the erratic data are less useful for evaluating risk.

In the hands of untrained students our test system has proved to be reliable and easy. Some modifications have been introduced during the last three years. Instead of intensely homogenizing the sediment prior to composing of the MES and to improve the recovery from the combustion of the sediment for radioanalysis, the extracted sediment is now homogenized with a mechanical mortar, without loss of recovery or replicability.

The use of centrifuge tubes with ground glass joints was found to be impractical because of the leakage during incubation caused by the ground part being fouled by "aufwuchs".

If compounds that are less soluble in water have to be tested, the solubility might be increased several fold by using MES water. The use of organic solvents for dosing, or of organic nutrients during incubation, is not recommended because of the unpredictable effect on the degradation of xenobiotics.[9-12] The use of relevant formulation ingredients that simulate the application of pesticides in field situation seems more realistic. The influence of the formulation on the behaviour and effects of pesticides in aquatic systems may be very large.[13]

Tenax[R], XAD-resins, activated charcoal or other modified silicas can be substituted for the SEP PAK[R] $C_{18}$ cartridge or the Baker SPE disposable columns. If these adsorbents are packed in small interchangeable columns suitable for HPLC, most of the analytical procedure can easily be automated.

## Acknowledgements

Special thanks are expressed to Bayer A.G., Leverkusen, F.R.G. for providing samples of parathion and aminoparathion, to Duphar B.V., Weesp, The Netherlands, for the use of the sample oxidizer and to the students Gerrit Bor and Rob Schut for their part in the experimental work.

## References

1. J. M. Giddings and G. K. Eddlemon, in Microcosms as Potential Screening Tools for Evaluating Transport and Effects of Toxic Substances, (U.S. Environmental Protection Agency, Washington, DC, 1980, EPA-600/3-80-042), pp. 15–170.
2. D. A. Graetz, G. Chesters, T. C. Daniel, L. W. Newland and G. B. Lee, *J. Water Pollut. Contr. Fed.* **42,** Part 2 R76 (1970).
3. B. M. Zuckerman, K. Deubert, M. Mackiewicz and H. Gunner, *Plant and Soil* **33,** 273 (1970).
4. A. V. Rao and N. Sethunathan, *Archs. Microbiol.* **97,** 203 (1974).
5. M. S. Sharom, J. R. W. Miles, C. R. Harris and F. L. McEwen, *Water Res.* **14,** 1089 (1980).
6. J. Katan and E. P. Lichenstein, *J. Agric. Food Chem.* **25,** 1404 (1977).
7. S. C. Schimmel, R. L. Garnas, J. M. Patrick, Jr. and J. C. Moore, *J. Agric. Food Chem.* **31,** 104 (1983).
8. A. W. Bourquin, M. A. Hood and R. L. Garnas, *Developm. Industr. Microbiol.* **18,** 185 (1977).

9. M. S. Sharom and J. R. W. Miles, *J. Environm. Sci. Health* **B16,** 703 (1981).
10. K. P. Rajaram and N. Sethunathan, *Soil Sci.* **119,** 296 (1975).
11. G. W. Gorder and E. P. Lichtenstein, *Can. J. Microbiol.* **26,** 475 (1980).
12. G. Schmidt, *Schr. Reihe Ver. Wass.-Boden-Lufthyg. Berlin-Dahlem* **46,** 156 (1975).
13. C. H. Schaefer and E. F. Dupras, Jr., *J. Agric. Food Chem.* **24,** 733 (1976).

# PART II

# ANALYTICAL ASPECTS

# Analytical Methodology for Organophosphorus Pesticides Used in Canada†

JAMES F. LAWRENCE

*Food Research Division, Bureau of Chemical Safety, Health Protection Branch, Ottawa, Ontario Canada K1A 0L2*

(*Received August 17, 1986; and in final form October 15, 1986*)

An overview of analytical methodology for the determination of organophosphate pesticides residues in foods is presented. Sample extraction is carried out with acetone followed by a dichloromethane–hexane partition. The organic extract is purified by automated gel permeation chromatography and analysed by capillary gas chromatography with flame photometric or thermionic detection. Confirmation can be carried out by a variety of chemical derivatization techniques including hydrolysis followed by reaction of the phosphate or phenol moiety, direct alkylation or trifluoroacetylation. Thin-layer chromatography with enzyme inhibition detection can be used as a rapid screening technique or to confirm results obtained by gas chromatography. Liquid chromatography has not been used much for the determination of organophosphorus compounds in foods.

KEY WORDS: Organophosphorus pesticides, insecticides, capillary gas chromatography, thin-layer chromatography, chemical derivatization, solvent extraction, gel permeation, flame photometric detector, thermionic detector.

## INFORMATION

In Canada there are some thirty-eight organophosphorus pesticides registered for use on food crops or food producing animals. They are

---

†Presented at the Workshop on Chemistry and Fate of Organophosphorus compounds, Amsterdam, June 18–20, 1986.

This article was first published in *International Journal of Environmental Analytical Chemistry*, Volume 29, Number 4 (1987).

J. F. LAWRENCE

registered individually for specific uses and cannot be used on crops for which they are not registered. Table I lists twenty-two compounds registered for use on fruits and vegetables with specific maximum residue limits. The limits are determined by taking into account the toxicity of the organophosphate as well as the rate and quantity of consumption of the food item. Table II lists those compounds registered for food crops on a "negligible residue" basis only. For analytical purposes this can be considered as those pesticides which yield residues below 0.1 $\mu$g/g in the crop at harvest. A number of organophosphates listed in Table I are also permitted for use on some crops on a negligible residue basis. Table III lists those organophosphates which are registered for use on animals and their maximum residue limits in the meat or meat by-products. In

**Table I**  Organophosphorus pesticides registered for use on fruits and vegetables

| Compound | Max. residue limit |
|---|---|
| Azinphos-methyl[a] (Guthion) | 0.2–5.0 ppm[b] |
| Bromophos | 1.5 |
| Carbophenothion (Trithion) | 0.5–0.8 |
| Demeton (Systox) | 0.2–0.75 |
| Diazinon[a] | 0.2–0.75 |
| Dichlorovos[a] (Vapona) | 0.25–2.0 |
| Dimethoate[a] (Cygon) | 0.5–2.0 |
| Dioxathion (Delnav) | 2.0–5.0 |
| Disulfoton[a] (Disyston) | 0.2–0.5 |
| Ethephon[a] (Etherel) | 0.5–20 |
| Ethion[a] | 0.5–2.5 |
| Malathion[a] | 0.5–8.0 |
| Methamidophos[a] (monitor) | 0.5–1.0 |
| Methidathion[a] | 0.2–2.0 |
| Mevinphos[a] (Phosdrin) | 0.2–0.25 |
| Monocrotophos (Azodrin) | 0.5–1.0 |
| Naled (Dibrom) | 0.5–3.0 |
| Parathion | 0.7–1.0 |
| Phosalone[a] (Zolone) | 1.5–15.0 |
| Phosmet[a] (Imidan) | 1.0–10.0 |
| Tetrachlorvinphos (Gardona) | 10.0 |
| Tetradifon (Tedion) | 1.0–100.0 |

[a]Also registered on a negligible residue basis.
[b]ppm = $\mu$g/g.

**Table II**  Organophosphorus pesticides registered for use on a negligible residue basis only

| Compound |
| --- |
| Chlorfenvinphos |
| Crotoxyphos |
| 2,4-DEP |
| Fensulfothion |
| Fenthion |
| Fonofos |
| Glyphosate |
| Isofenphos |
| Menazon |
| Oxy-demeton-methyl |
| Phorate |
| Terbuphos |

**Table III**  Organophosphorus compounds registered for use on animals

| Compound | Max. residue limit (meat and meat by-products) |
| --- | --- |
| Chlorpyrifos (Dursban) | 1.0 ppm |
| Coumaphos (Co-Ral) | 0.5 ppm |
| Crufomate (Ruelene) | 1.0 ppm |
| Dioxathion | 1.0 ppm |
| Ethion | 2.5 ppm |
| Ronnel | 3–7.5 ppm |
| Tetrachlorvinphos (Gardona) | 0.75–1.5 ppm |

addition to those organophosphates registered in Canada we need to have methodology for those compounds which may appear as residues in imported foods. The total lists of organophosphorus compounds which must be considered is over one hundred.

## SAMPLE PREPARATION

The main routine methodology for organophosphate determinations at the Canadian Department of Health and Welfare presently includes capillary gas chromatography with phosphorus selective detection with either flame photometric or thermionic detectors. Sample cleanup is predominantly done by gel permeation[1] (size exclusion) chromatography. Figure 1 shows an overall schematic of the procedure for fruits and vegetables. It is designed to fit into our general multi-pesticide residue screening methodology. Acetone is used as a universial extractant and this is partitioned with dichloromethane/hexane, then the organic extracts are evaporated to a small volume and dissolved in acetone or dichloromethane/cyclohexane (1 + 1) for gel permeation cleanup for the determination of the

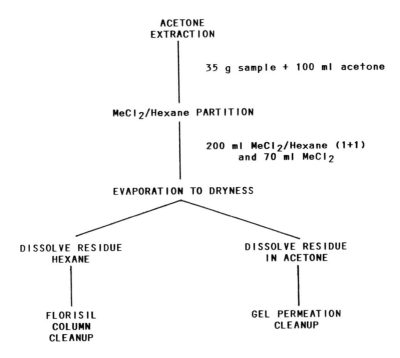

**Figure 1** Schematic for sample extraction and cleanup for organophosphate insecticides in fruits and vegetables.

organophosphorus compounds. The extracts are further cleaned by Florisil or other type of open column chromatography for the determination of other classes of pesticides. Fat samples are directly diluted with acetone or dichloromethane/cyclohexane for gel permeation cleanup.

The gel permeation apparatus (Autoprep 1002, ABC Labs) is automated and capable of handling 23 samples in a single run. The system normally employs 60 g of Biobeads SX-3 with eluting solvents of either dichloromethane/cyclohexane $(1 + 1)$ as mentioned above or dichloromethane/acetone $(7 + 3)$ at a flowrate of 5 mL/min. Table IV lists the approximate elution volumes for a variety of substances. It can be seen that pesticides including organophosphates, elute in about 150–200 mL while lipid material such as fish oil elutes in less than 100 mL. The main function of gel permeation is to remove the higher molecular weight lipid material which so often interferes in gas chromatographic analyses of pesticides. After this cleanup the extracts are generally clean enough to be concentrated and analysed directly by gas chromatography with selective detection. Table V shows the recoveries obtained for fourteen organophosphates spiked in vegetable oil and lettuce at levels of 0.1–0.25 ppm ($\mu$g/g). It can be seen that with the exception of fenthion and perhaps ethion, recoveries are very good. It is possible that the elution pattern for these two pesticides is different from the rest resulting in losses.

**Table IV**  Approximate elution volumes of compounds from Biobeads SX-3 with dichloromethane/cyclohexane $(1 + 1)$

| Substance | Elution volume (mL) |
| --- | --- |
| Fish oil | 40–100 |
| Aliphatics | 100–160 |
| Stearic acid | 110–150 |
| Phthalates | 115–150 |
| Pesticides | 150–200 |
| PCB's | 170–210 |
| Phenols | 175–240 |
| Polycyclic aromatics | 190–260 |
| Nitrophenols | 240–315 |

**Table V**   Organophosphates evaluated for GPC cleanup

| Compound | Veg. oil[a] | | Lettuce[b] | |
|---|---|---|---|---|
| | Level added (ppm) | Recovery (%) | Level added (ppm) | Recovery (%) |
| Diazinon | 0.25 | 82 | 0.10 | 96 |
| Parathion | 0.25 | 95 | 0.10 | 88 |
| Ethion | 0.25 | 68 | 0.10 | 91 |
| Fenthion | 0.10 | 46 | 0.10 | 109 |
| Ronnel | 0.10 | 107 | — | — |
| Malathion | 0.15 | 100 | — | — |
| Chlorpyrifos | — | — | 0.10 | 110 |
| Carbophenothion | — | — | 0.10 | 99 |
| Dimethoate | — | — | 0.10 | 79 |
| Dimethoxon | — | — | 0.10 | 94 |
| Fonofos | — | — | 0.10 | 103 |
| Methamidophos | — | — | 0.10 | 81 |
| Fensulfothion | — | — | 0.10 | 104 |
| Phosphamidon | — | — | 0.10 | 106 |

[a]Veg. oil: $MeCl_2$/cyclohexane $(1 + 1)$.
[b]Lettuce: $MeCl_2$/acetone $(7 + 3)$.

## GAS CHROMATOGRAPHY

Figure 2 shows a capillary gas chromatographic separation of 38 organophosphorus pesticides and related compounds with a DB-17 column and flame-photometric detection. This detector was found to be generally more selective than the thermionic detector. Figures 3 and 4 compare the two detectors for extracts of green peppers (containing 0.37 ppm methamidophos and 1.3 ppm acephate) and oranges (containing approximately 0.1 ppm ethion oxon and 1.0 ppm ethion). In both cases the flame photometric results show fewer peaks than the thermionic ones, although both detectors easily made the determination of the indicated organophosphates possible. However, in monitoring unknown samples for a large number of organophosphates, the flame photometric detector would be preferred because of the fewer peaks observed. In the case of the thermionic detector, the additional peaks observed would have to be matched to the mixed standards (as shown in Figure 2, for example)

107

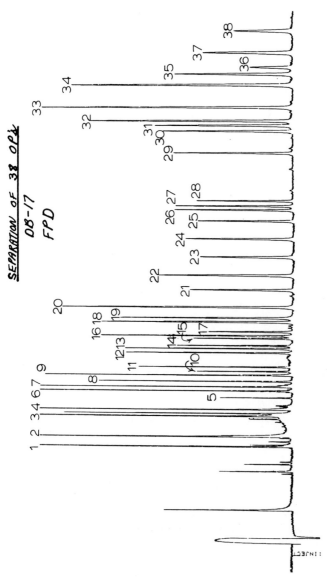

**Figure 2** Chromatogram of 38 organophosphates on a DB-17 column with flame-photometric detection. 1. trichlorofon, 2. methamidophos, 3. phosdrin, 4. acephate, 5. phorate oxon, 6. phorate, 7. dimethoate oxon, 8. diazinon, 9. monocrotophos, 10. dioxathion, 11. dimethoate, 12. methyl chlorpyrifos, 13. malaoxon, 14. methyl parathion, 15. chlorpyrifos, 16. malathion, 17. parathion, 18. fenthion, 19. ruelene, 20. chlorfenvinphos, 21. phenthoate, 22. tetrachlorvinphos, 23. methadithion, 24. ethion oxon, 25. ethion, 26. sulprofos, 27. carbophenothion, 28. fensulfothion, 29. EPN, 30. leptophos, 31. phosolone, 32. phosmet, 33, azinphos methyl oxon, 34. menazon, 35. azinphos methyl, 36. dialifos, 37. azinphos ethyl, 38, coumaphos.

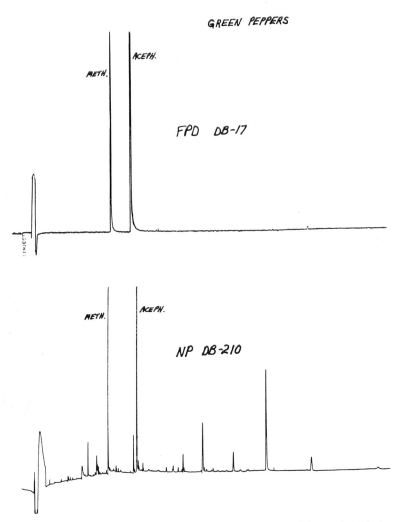

**Figure 3** Chromatograms of an extract of green pepper containing methamidophos and acephate with flame photometric (FPD) and thermionic (NP) detection using a DB-17 and a DB-210 column respectively.

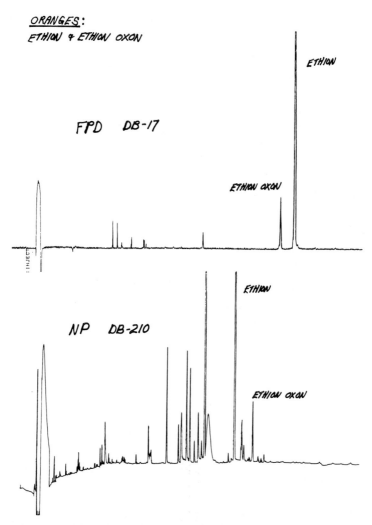

**Figure 4** Chromatogram of an extract of oranges containing ethion and ethion oxon with flame photometric (FPD) and thermionic (NP) detection using a DB-17 and a DB-210 column respectively.

J. F. LAWRENCE

for confirmation of identity. This takes time and there is a possibility that a wrong identification could be made. In order to avoid this, confirmation of the peaks by some other means (different columns or detectors, or chemical derivatization) would need to be carried out.

## CHEMICAL DERIVATIZATION

Chemical derivatization is a very useful means for confirming the identity of an unknown peak. By carrying out a chemical reaction on an unknown and matching the chromatographic properties of the product to that obtained from authentic material, an analyst can obtain useful information to help unequivocally identify the unknown. Figures 5 and 6 show two approaches for forming derivatives of parathion. In Figure 5 parathion is hydrolysed under basic conditions to yield the thiophosphate moiety and p-nitrophenol.[2] The phenol is

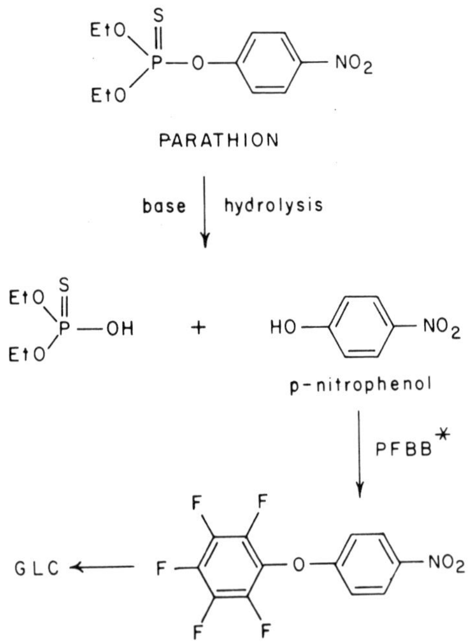

**Figure 5** Hydrolysis of parathion and derivatization with pentafluorobenzyl bromide (PFBB).

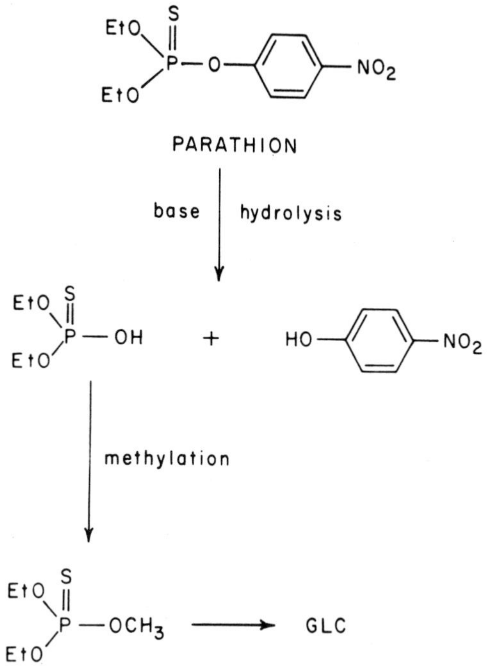

Figure 6    Hydrolysis of parathion and derivatization with diazomethane.

then derivatized with pentafluorobenzyl bromide (PFBB) to form the ether derivative which is very sensitive to electron capture detection. In Figure 6 the same hydrolysis procedure is used, however the phosphorus moiety is methylated for determination by gas chromatography with flame photometric detection.[3] In most cases a single step derivatization is preferred but can only be carried out when the pesticide structure permits it. Figure 7 shows the direct alkylation of dimethoate using methyl iodide with sodium hydride as a base.[4] This reaction simply adds a methyl group to the molecule which results in a retention time shift. Since the phosphorus atom is still present in the derivative, the selective flame photometric detector can still be used for detection.

In some cases basic conditions cannot be used. For example although Azodrin (monocrotophos) has a free N-H proton similar to dimethoate, the methyl iodide/sodium hybride alkylation reaction

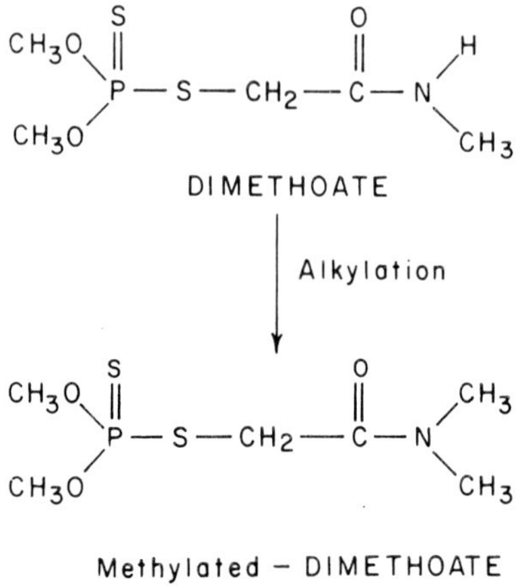

DIMETHOATE

Alkylation

Methylated — DIMETHOATE

Figure 7    Alkylation of dimethoate with methyl iodide/sodium hydride.

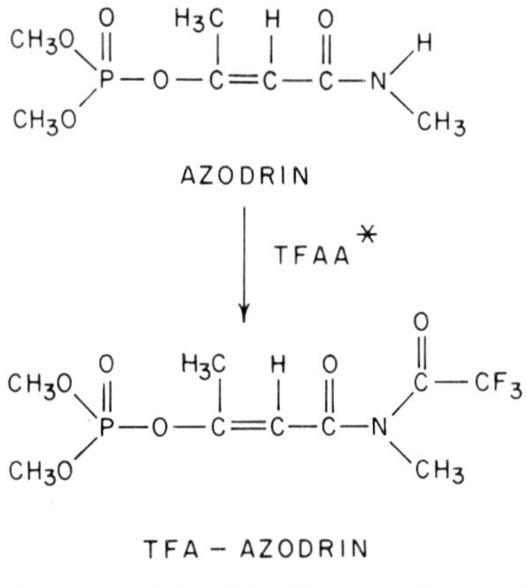

AZODRIN

TFAA *

TFA — AZODRIN

Figure 8    Reaction of Azodrin with trifluoroacetic anhydride (TFAA).

was not successful due to decomposition of the insecticide under the conditions used. However, if acidic conditions such as reaction with trifluoroacetic anhydride (TFAA) are used, a well defined product can be obtained.[5] Figure 8 illustrates the reaction scheme using TFAA. The product is more volatile than the parent so elutes earlier under the same conditions. Figure 9 shows examples of the detection

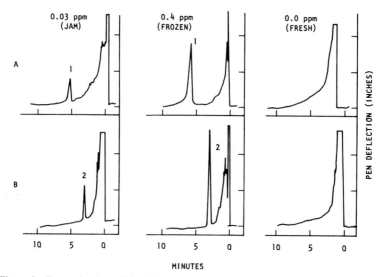

**Figure 9**   Determination of Azodrin in fresh and frozen strawberries and strawberry jam. A. Upper chromatograms, direct analyses. B. Lower chromatograms, after TFAA reaction. 1 = azodrin. 2 = TFA-azodrin.

and confirmation of Azodrin in strawberries and strawberry jam using packed column gas chromatography and flame photometric detection.

## THIN LAYER CHROMATOGRAPHY

Thin-layer chromatography has also been used both as a means of confirmation of organophosphate identity and as a rapid screening technique.[6-8] The detection method employed is enzyme inhibition normally using a spray solution consisting of mixtures of beef liver esterases and the substrate, 5-bromoindoxyl acetate. The approach is very selective as well as sensitive, detecting many organophosphates at low nanogram levels. Compounds such as azinphos-methyl, carbophenothion, diazinon, ethion, malathion, mevinphos, parathion and others have been detected in spiked extracts of fruits and vegetables at levels of 0.1–0.2 ppm. Figure 10 shows a thin-layer

A Diagram of the Relative Mobility of Nineteen Pesticides on a Silica gel
TLC Plate Developed in 20% Acetone in Hexane

1 - phorate (Thimet)
2 - dichlofenthion
3 - disulfoton
4 - ronnel
5 - tetradifon
6 - ethion
7 - diazinon
8 - parathion
9 - methyl parathion
10 - Sumithion
11 - malathion
12 - coumaphos
13 - demeton
14 - naled
15 - azinphosmethyl
16 - carbaryl
17 - phosmet
18 - crufomate
19 - mevinphos

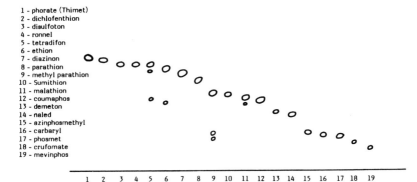

    1   2   3   4   5   6   7   8   9   10   11   12   13   14   15   16   17  18  19

**Figure 10**   Thin-layer chromatogram of 18 organophosphates and carbaryl. Small extra spots represent impurities.

separation of 18 organophosphates and carbaryl which are detectable using the enzyme inhibition technique.

## LIQUID CHROMATOGRAPHY

Although, liquid chromatography has been evaluated for some organophosphates either directly or as derivatives,[9,10] the method has not found widespread use for routine determination. This is mainly because capillary gas chromatography with flame photometric (or thermionic) detection is ideally suited for the direct quantitation of most organophosphates in foods at levels down to the low ppb (ng/g) range with minimal sample cleanup.

### References

1. *Analytical Methods for Pesticide Residues in Foods* (H. McLeod and R. Graham, ed.) (Health and Welfare Canada, Ottawa, 1986).
2. J. A. Coburn and A. S. Y. Chau, *J. Assoc. Offic. Anal. Chem.* **57,** 1212 (1974).
3. M. T. Shafik and H. F. Enos, *J. Agric. Food Chem.* **17,** 1186 (1969).
4. R. Greenhalgh and J. Kovacicova, *J. Agric. Food Chem.* **23,** 325 (1975).
5. J. F. Lawrence and H. A. McLeod, *J. Assoc. Offic. Anal. Chem.* **59,** 639 (1976).

6. C. E. Mendoza, *Res. Rev.* **43,** 105 (1972).
7. A. M. Gardiner, *J. Assoc. Offic. Anal. Chem.* **54,** 517 (1971).
8. C. E. Mendoza. In: *Pesticide Analysis* (K. G. Das, ed.) (M. Dekker Inc., New York, 1981), p. 1.
9. J. F. Lawrence, *Analytical Methods for Pesticides and Plant Growth Regulators Volume 12* (G. Zweig and J. Sherma, ed.) (Academic Press Inc., New York, 1982).
10. J. F. Lawrence, C. Renault and R. W. Frei, *J. Chromatogr.* **121,** 343 (1976).

# Various Means for the Detection of Organophosphorous Compounds

## M. S. NIEUWENHUIZEN and A. W. BARENDSZ

*Prins Maurits Laboratory TNO, P.O. Box 45, 2280 AA Rijswijk, The Netherlands*

(*Received June 16, 1986; in final form September 10, 1986*)

Several reaction principles can be used for the detection of organophosphorous compounds. E.g. the "molybdenum blue" method, the colorimetric and fluorimetric variants of the Schoenemann reaction, electrochemical detection of cyanides upon reaction with certain oximes, photoemission of exitated HPO molecules in a hydrogen rich flame and the well known enzymatic detection of cholinesterase inhibiting compounds.

Various detection systems have been developed varying from simple, manually operated devices to more sophisticated and continuous functioning detectors.

These well established detection principles and some new emerging techniques, like microsensors using the surface acoustic wave technique, will be discussed.

## 1. INTRODUCTION

The increasing use of toxic (thio)phosphonates, phosphates and carbamates as pesticides, insecticides, corrosion inhibitors and anti-knock compounds, as well as the potential use of organophosphorous compounds as nerve agents have necessitated the development of highly sensitive and fast reacting detector devices. Before World War II the presence of chemical warfare agents was to be detected with

---

†Presented at the Workshop on Chemistry and Fate of Organophosphorous Compounds, Amsterdam, Holland, June 18–20, 1986.

This article was first published in *International Journal of Environmental Analytical Chemistry*, Volume 29, Numbers 1 + 2 (1987).

biological indicators (e.g. birds were taken to the battlefront) or by their olfactory properties. The days have passed that these methods can be employed without causing any harm. In the Netherlands, especially at the Prins Maurits Laboratory TNO, various systems have been developed for the detection of chemical warfare agents, that are of paramount importance as an integral part of a total defence system against chemical warfare. In this paper some past, present and future principles of detection will be reviewed.[1,2]

The organophosphorous compounds considered can be described with the general formula:

$$R_1 \diagdown \quad \diagup O$$
$$P$$
$$R_2 \diagup \quad \diagdown X$$

in which $R_1$ and $R_2$ are alkyl, alkoxy or iminogroups and X is a leaving group. Well-known examples of nerve agents are tabun, sarin and soman (G-agents) and VX (V-agent). Their structure is given in Table I.

**Table I**    The structure of some nerve agents

| Nerve agent | $R_1$ | $R_2$ | X |
|-------------|-------|-------|---|
| tabun | $N(Me)_2$ | Et | CN |
| sarin | $OCH(Me)_2$ | Me | F |
| soman | $OCH(Me)C(Me)_3$ | Me | F |
| VX | OEt | Me | $SC_2H_4N(iPr)_2$ |

During the impact of nerve agents on the human body the active centre of the enzyme acetylcholinesterase is irreversibly phosphonyl-ated. Once the enzyme is deactivated it can not fulfil its function anymore viz. the hydrolysis of acetylcholine, which plays an important role in the transfer of nerve stimuli. Consequently, the acetylcholine concentration is increased and the specific symptoms of poisoning appear.

## 2. PAST AND PRESENT DETECTION TECHNIQUES

Several methods of detection will be described in this section and examples of their applications will be given. However, this will be limited to chemical and biochemical methods.

### The "molybdenum blue" method

One of the oldest chemical methods for the detection of organophosphorous compounds is an extension of an ordinary phosphate detection method.[3] The organophosphorous compounds are destroyed oxidatively to orthophosphate by means of e.g. sulphuric acid or perchloric acid (Reaction scheme 1).

$$
\begin{array}{c}
R_1 \\[-2pt]
\phantom{R_1}\diagdown \quad \diagup\!\!\diagup O \\[-4pt]
\quad\quad P \\[-4pt]
\phantom{R_2}\diagup \quad \diagdown \\[-2pt]
R_2 \quad\quad X
\end{array}
\longrightarrow PO_4^{3-}
$$

$$PO_4^{3-} + (NH_4)_6Mo_7O_{24} \longrightarrow H_3\left[P(Mo_3O_{10})_4\right] \xrightarrow{\text{Reductor}} \text{Blue colour}$$

**Reaction scheme 1**    The molybdenum blue method.

The orthophosphate anion reacts with ammonium heptamolybdate in sulphuric acid to form a reddish phosphomolybdenic acid, which is then reduced to a blue coloured compound. As this method suffers from many interferences West et al.[4,5] modified it by adding o-dianisidine in the last step thus improving its sensitivity and selectivity. At the moment the molybdenum blue method is hardly in use anymore for the detection of organophosphorous compounds. A similar method for the detection of arsenic compounds is still being used.

### The Schoenemann reaction

The Schoenemann reaction[6] is based on the fact that peroxophosphonates do oxidize amines much easier than other peroxy ions (Reaction scheme 2).

$$H_2O_2 + OH^- \longrightarrow HOO^- + H_2O$$

**Reaction scheme 2**   The Schoenemann reaction.

First, the organophosphorous compound reacts with hydrogen peroxide or sodium borate in an alkaline solution (pH 9–10). Then the peroxophosphonate reacts with a leuco-dye like benzidine or o-dianisidine to form an orange-brown coloured product. In certain modified reactions the amine is replaced by a precursor of a chemiluminescent compound (e.g. luminol)[7] or a fluorescent compound (e.g. indole).[8] Many organophosphorous compounds have been tested and it was shown that only labile P-X bonds (G-agents) give a positive reaction. Other ions that may be present or hydrolysis products of phosphonates do not interfere, and sometimes even enhance the luminescence.

The luminescent variant of the Schoenemann reaction finds practical application for civil and military purposes in portable, mobile and stationary laboratories.[9,10] The colorimetric variant of the Schoenemann reaction has been applied in so-called detection tubes. Air passes through a silica layer containing the dye. Then the peroxide solution is added from an internal breakable ampoule. Draeger (FRG) marketed such a detection tube for tabun and sarin (type Tabun-Sarin) having a detection limit of about 10 mg/m3. It should be noted that these tubes have been replaced by detection tubes employing the enzymatic detection principle (type Phosphorsaureester) instead of the Schoenemann reaction.

## Electrochemical detection

An electrochemical method of detection is based on the fact that some organophosphorous compounds react with certain oximes like isonitrosobenzoyl acetone (IBA)[11] (Reaction scheme 3).

**Reaction scheme 3**   Reaction of organophosphorous compounds and oximes.

The anion of the oxime reacts with the organophosphorous compound having a good leaving group X. In case of V-agents a reactive leaving group is introduced upon reaction with a silver-fluoride conversion filter. The intermediates yield cyanide ions via a Beckmann rearrangement in an alkaline solution. Upon reaction with IBA one cyanide ion yields two cyanide ions in return. The cyanide can be detected via a colorimetric reaction with p-nitrobenzaldehyde or can be determined electrochemically.

Based on this electrochemical method of detection a miniature chemical agent detector (ICAD) has been developed by Bendix (USA). The detector consists of a reusable electronic module (processor, audible alarm and warning light), and a disposable sensor module containing the battery power source and the sensor cells.

## Photometric detection

Normally there are no organophosphorous compounds in the atmosphere. Therefore a more or less selective method could be developed based on the detection of the phosphorus atom using flame emission.[12] When organophosphorous compounds are burnt in a hydrogen-rich flame excited HPO molecules are formed (Reaction scheme 4).

$$R_1 \diagdown \quad \diagup\!\!\diagup O \qquad \qquad$$
$$\qquad P \qquad \xrightarrow[\Delta T]{H_2} \quad HPO^* \longrightarrow HPO + h\nu$$
$$R_2 \diagup \quad \diagdown X$$

**Reaction scheme 4**    Photometric detection.

When these excited molecules fall back to their ground state light is emitted in the range of 500–550 nm. Especially, the intensive band near 526 nm is characteristic for phosphorus and can be measured. In France an automatic warning system (Detalac) has been developed for military purposes. The hydrogen is generated in a disposable unit by reaction of aluminum and hydrogen fluoride. For civil use both a GC detector (Tracor) and an environmental monitor (Meloy) have been developed able to measure both phosphorus as well as sulphur.

## Detection based on enzyme inhibition effects

By far the most important method of detection at this moment is based on the enzyme inhibiting properties of certain organophosphorous compounds.[13] As such these detection devices can be regarded as first examples of biosensors. In these devices the naturally involved acetylcholine is replaced by a chromogenic substrate, like 2,6-dichloroindophenyl acetate (Reaction scheme 5).

Normally, the enzyme will catalyze the hydrolysis of the ester group of the substrate, which causes a distinct colour change from orange-red to blue. The difference in colour between the active and inhibited state of the enzyme can be observed visually or spectrophotometrically.

**Reaction scheme 5**   The colorimetric version of the enzymatic detection reaction.

Based on this principle a number of simple, manually operated field tests (detection tubes, reagent papers) as well as automatic functioning monitors (warning equipment) have been developed. In the Netherlands the "button"-detector is marketed by Duphar BV. This coin-sized individual detector (Figure 1) consists of a plastic holder containing two separated air-permeable reagent papers, one impregnated with the enzyme and silica and the other with the substrate. Also, the detector contains a reservoir which releases the reagent solution when punctured. When air is drawn through the enzyme paper, the organophosphorous compounds are adsorbed on the silica. Then the counterparts of the "button"-detector are pressed together. The reagent solution is released, the reagent papers are wetted and at the same time pressed together initiating the enzymatic detection reaction. After two minutes the blue colour of the decomposed substrate can be clearly observed.

A similar test (water detection sticks) is available for checking the presence of cholinesterase inhibiting compounds in waste water streams. This test is produced by Rijling B.V. in the Netherlands. In automatic functioning monitors (designed as warning equipment) the residual activity of the enzyme can be determined either by a colorimetric reaction or electrochemically. In the Netherlands (Oldelft) the enzymatic reaction is performed on a continuous moving tape. The colour change is observed photometrically, followed by a conversion into audible or visible warning signals. In England (EMI Thorn) butylthiocholine is used as substrate. When it is decomposed by the immobilized enzyme the produced thiocholine

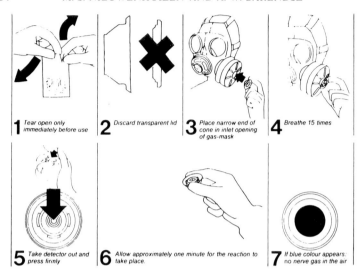

| | | | |
|---|---|---|---|
| **1** Tear open only immediately before use | **2** Discard transparent lid | **3** Place narrow end of cone in inlet opening of gas-mask | **4** Breathe 15 times |
| **5** Take detector out and press firmly | **6** Allow approximately one minute for the reaction to take place. | | **7** If blue colour appears: no nerve gas in the air |

**Figure 1** The "button" enzymatic chemical nerve agents detector.

is measured electrochemically by anodic oxidation to form choline-disulphide (Reaction scheme 6).

$$R \ S \ \overset{\overset{\displaystyle O}{\|}}{C} \ R' \xrightarrow{\text{Enzyme}} RSH + R'COOH$$

$$RSH + OH^- \longrightarrow RS^- + H_2O$$

$$2RS^- \rightleftharpoons RSSR + 2e$$

**Reaction scheme 6** The electrochemical version of the enzymatic detection reaction.

## Ion mobility detection

This detection technique is based on measuring the difference in mobility of various ions under influence of an electric field.[14,15] In the measuring cell an air stream passes a radioactive source (e.g.

63-Ni) generating ions like $H_3O^+$ or $H_2O_3^-$. Multiple collisions of these ions with the organophosphorous compounds cause the formation of charged ion clusters. Especially organophosphorous compounds are extremely efficient in forming such clusters. Also these clusters are heavier and therefore more stable than the ion clusters of other compounds normally occuring in the atmosphere. The charged clusters are now brought into a path with an electric potential gradient or an extremely tortuous path. The ions are separated by their difference in mobility (weight).

This technique is being applied in a number of alarm systems for nerve agents. In the USA the M43A1 chemical agent detector (Honeywell) uses a tortuous path. In England a portable and handheld chemical agent monitor (CAM) is developed (Graseby Dynamics) using the time of flight separator.

## 3. FUTURE TRENDS IN DETECTION METHODS

### Microsensors

In modern warfare the units tend to operate more dispersed. Individual detection devices will then become more and more important, requiring further automation and miniaturization of detection equipment, i.e. microsensors are to be developed.

The objectives to be met by the microsensors will be: small size, light weight, robust and low cost. With respect to the general performance characteristics a sensitive (1–10 ppb), fast responding (0–10 s), reversible and selective response of the sensor will be required. The detection principle must be readily adaptable to various toxic compounds and the equipment must be simple to operate and have a long shelf life (5–10 years).

In recent years many miniaturizing technologies have been investigated. Chemical Sensitive Semiconductor Devices (CSSDs) such as chemiresistors and CHEMFETs, optical wave guides and acoustic sensors.[16] King[17] introduced the use of bulk acoustic waves (BAW) with his Piezoelectric Sorption Detector. Wohltjen and Dessy[18] extended this detection principle to surface acoustic waves (SAW).

### Surface acoustic wave devices

Acoustic waves can be generated by applying an interdigital transducer on a piezoelectric substrate. When the crystal orientation of the substrate is properly chosen these waves propagate along the surface to the other transducer. Changes in the physical characteristics of the wave path will cause changes in the propagation velocity, which can be determined very sensitively and accurately by frequency measurement. When the wave path is covered with a so-called chemical interface (Figure 2) which reacts selectively and reversibly with the gas to be measured a SAW chemosensor can be realized.

In our laboratory a SAW-chemosensor for NO2 has been developed using metal-free and metallophthalocyanines as chemical interface.[19,20] A so-called dual delay-line configuration was used in order to compensate for non-specific effects. (Figure 2).

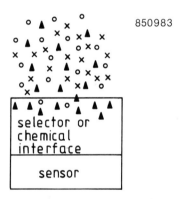

850983

**Figure 2**　Schematic presentation of a chemosensor, being a combination of a sensor and a selector or chemical interface.

The type of interaction between the gas and the chemical interface, the structure (morphology) of the interface material and the way it is attached to the surface of the sensor will determine the ultimate performance characteristics of the detection device.

When using the *absorption* phenomenon selectivity is determined only by slight differences in the partition coefficient between gas and chemical interface. Generally a poor selective sensor is obtained. In case of *adsorption* only weak interactions occur when the gas deposits on the interface material. The energies involved range from van der Waals' forces (0–10 kJ/mole) to acid-base interactions

($< 40\,$kJ/mole). In case of *chemisorption* very strong interactions occur at the chemical interface. Chemical bonds are broken and other covalent or ionogenic bonds can be formed (energy per bond $\sim 300\,$kJ/mole). As far as selectivity is concerned a specific chemical reaction or chemisorption is preferred. However, the gases are then strongly and often irreversibly bonded. So, for reversibility the weaker adsorptive interactions are preferred. Obviously these conflicting requirements need a compromise. Such a compromise can be found in the area of *coordination chemistry* or charge transfer complex formation (Figure 3).

850984

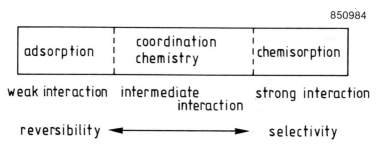

**Figure 3**  Gas-solid interactions with respect to the selectivity and reversibility as some general performance characteristics.

In Table II the compounds are listed which have been suggested for the use as chemical interface. A division is made with respect to the expected physical or chemical interaction. It should be emphasized that these classifications are sometimes rather arbitrary. Often literature is not precise with regard to the information about the chemical interactions involved.

It can be concluded that three types of interaction could give rise to a more or less selective detection of organophosphorous compounds:

—Compounds containing metal ions like Cu(II) or Fe(III). Here coordination compounds are formed between the metal ion and the organophosphorous compounds. It was shown already that coordination compounds may act as a compromise between the weak and reversible adsorption and the strong and often irreversible chemical reaction.

**Table II**   Types of chemical interfaces found in literature for sensors using gas/solid interactions

| Interaction/ chemical interface | Sensor | Organo P compounds (1) | Ref. |
|---|---|---|---|
| *Absorption* | | | |
| poly(vinylpyrrolidone) | BAW | DIMP | 21 |
| *Chemisorption* | | | |
| oxime coatings | BAW/electrode | (2), DDVP | 2,22,23 |
| 3-PAD (1) | BAW | DIMP, pesticides | 24 |
| 3-PAD + PVP (1) | BAW | DIMP | 21 |
| *Coordination* | | | |
| poly-amino Cu-comp. | CHEMFET/resistor | DIMP, DMMP | 25–28 |
| Fe(III)-chloride | BAW | DIMP | 29 |
| Au-, Ag-, Ni-compounds | BAW | DIMP | 30 |
| Cu(II)-PC (1) | resistor | DMMP | 21 |
| Cu(II) TMEDA/PVP (1) | BAW | DIMP | 31 |
| nitro PPA (1) | resistor | DIMP | 32 |
| *Biosensing* | | | |
| enzyme (3) | IR | general | 33 |

(1) DIMP   = diisopropyl methylphosphonate
   DMMP   = dimethyl methyl phosphonate
   DDVP   = dimethyl 2,2-dichlorovinylphosphonate
   PC      = phthalocyanine
   3-PAD   = 1-n-octyl-3-(hydroxyiminomethyl) pyridinium iodide
   TMEDA  = tetramethylenediamine
   PVP     = poly(vinylpyrrolidone)
   PPA     = poly(phenylacetylene)
(2) all kinds of fluorophosphonates
(3) irreversible reaction

—Compounds containing oxime groups. However, they often react irreversibly with the organophosphorous compounds.

—Enzymes. They interact irreversibly with the organophosphorous compounds as well.

The common disadvantage of the examples mentioned in the literature is the way the chemical interfaces are bonded to the surface of the sensor i.e. physical attachment. Especially at elevated temperatures and in gas streams this method will cause serious stability problems. Therefore, we adhere to the concept of covalently bonded chemical interfaces on the surface of the chemosensor, which will lead to smaller response times and improved stability as well as a more efficient relation between mass changes and the wave propagation properties. The search for suitable chemical interfaces for the detection of organophosphorous compounds, which can be immobilized at the surface of the sensor is in progress at our laboratory.

## References

1. S. J. Smith, *Talanta* **30**, 725 (1983).
2. A. Snow, H. Wohltjen, N. Jarvis and D. Dominguez, NRL Memorandum Report 5050, Naval Research Laboratory, Washington 1982.
3. U. Bartels and H. Hayme, *Chem. Techn.* **11**, 156 (1959).
4. J. W. Robinson and P. W. West, *Microchem. J.* **1**, 93 (1957).
5. C. M. Welch and P. W. West, *Anal. Chem.* **29**, 874 (1957).
6. R. B. R. Schoenemann, "New reaction for the detection of metalloid-nonmetal linkages" DB 119877, Office of Publication Board U.S. Dept. of Commerce, 1944.
7. J. Goldenson, *Anal. Chem.* **29**, 877 (1957).
8. B. Gehauf and J. Goldenson, *Anal. Chem.* **29**, 276 (1957).
9. J. C. Young, J. R. Parsons and H. E. Reeber, *Anal. Chem.* **30**, 1236 (1958).
10. R. H. Cherry, G. M. Foley, C. O. Badgett, R. D. Eaton and H. D. Smith, *Anal. Chem.* **30**, 1239 (1958).
11. E. J. Poziomek, E. V. Crabtree and D. N. Kramer, *Microchem. J.* **18**, 622 (1973).
12. H. Frosting, *J. Phys. E.: Sci. Instr.* **6**, 863 (1978).
13. E. Boyer. In: *The Enzymes* Vol. 5A/B (E. Boyer, ed.) (Pergamon, New York, 1971).
14. E. W. McDavid and E. A. Mason, *The Mobility and Diffusion of Ions in Gases* (Wiley, New York, 1973).
15. J. M. Preston, F. W. Karasek and S. H. Kim, *Anal. Chem.* **49**, 1746 (1977).
16. J. E. Brignell and A. P. Dorey, *J. Phys. E.: Sci. Instrum.* **26**, 947 (1983).
17. W. H. King, Jr., *Anal. Chem.* **36**, 1735 (1964).
18. H. Wohltjen and R. Dessy, *Anal. Chem.* **51**, 1458, 1465, 1470 (1979).
19. A. W. Barendsz, J. C. Vis, M. S. Nieuwenhuizen, E. Nieuwkoop, M. J. Vellekoop, W. J. Ghijsen and A. Venema, *Proc. IEEE Ultrasonics Symposium*, San Francisco, 1985.
20. M. S. Nieuwenhuizen, A. W. Barendsz, E. Nieuwkoop, M. J. Vellekoop and A. Venema, *Electronics Letters* **22**, 184 (1986).
21. H. Wohltjen, *Proc. Int. Symp. Protection Against Chemical Warfare Agents*, Stockholm 51 (1983).
22. W. M. Shackelford and G. G. Guilbault, *Anal. Chim. Acta* **73**, 383 (1974).
23. G. Olofsson, *Proc. 3rd Int. Conf. on Sensors and Actuators*, Philadelphia 443 (1985).
24. Y. Tomita and G. G. Guilbault, *Anal. Chem.* **52**, 1484 (1984).
25. J. Janata, R. Huber, R. Cohen and E. S. Koleser, Report SAM-TR-80-25, (1980) Dept. of Bioengineering, Univ. of Utah.

26. J. Janata and E. S. Kolesar, Report SAM-TR-82-10 (1982), Dept. of Bioengineering, Univ. of Utah.
27. J. Janata and D. Gehmlich, Report USAFSAM-TR-83-47 (1983), Univ. of Utah.
28. G. G. Guilbault, J. Affolter, Y. Tomita and E. S. Kolesar, *Anal. Chem.* **53**, 2057 (1981).
29. E. P. Scheide and G. G. Guilbault, *Anal. Chem.* **44**, 1764 (1972).
30. G. Kristoff and G. G. Guilbault, *Anal. Chim. Acta* **149**, 337 (1983).
31. G. G. Guilbault, J. Kristoff and D. Owen, *Anal Chem.* **57**, 1754 (1985).
32. S. E. Wentworth and P. R. Bergquist, *J. Poly. Sci.: Polym. Chem. Ed.* **23**, 2197 (1985).
33. A. E. Grow, U.S. Patent 4411989 (1983).

# Evaluation of Combined Flow Injection–High Performance Liquid Chromatography for the Determination of Three Organophosphorus Pesticides in Liquid Wastes†

A. FARRAN and J. DE PABLO

*Department Quimica, ETSEIB (UPC), Diagonal 647, 08028 Barcelona, Spain*

(*Received August 17, 1986; in final form November 22, 1986*)

The separation and analysis of a mixture of diazinon, azinphos-methyl and fenthion have been studied by using HPLC with UV detector. In order to carry out this analysis in a simple and rapid way a system of flow injection (FIA) has been connected to the HPLC apparatus. This combined technique has shown great possibilities in the analysis of pesticides.

The removal of pesticides from liquid wastes has been studied by means of adsorption on activated carbon using this technique.

KEY WORDS: HPLC, FIA, flow injection analysis, organophosphorus pesticides, activated carbon, adsorption.

## INTRODUCTION

A recent study[1] concerning pesticide factories in Spain has indicated that the analysis and treatment of waste water from these factories is

---

†Presented at the Workshop on Chemistry and Fate of Organophosphorus Compounds, Amsterdam, Holland, June 18–20, 1986.

This article was first published in *International Journal of Environmental Analytical Chemistry*, Volume 30, Numbers 1 + 2 (1987).

not carried out very often. In this context, analysis and removal of organophosphorus pesticides from waste water are of great importance.

In this study three organophosphorus pesticides widely used in Spain:[1] diazinon, azinphos methyl and fenthion have been chosen.

One analytical technique especially suited to pesticide analysis is high performance liquid chromatography (HPLC) with UV absorbance detector.[2,3] In order to analyze these pesticides in a simple and rapid way a system of flow injection (FIA)[4] has been connected to the HPLC apparatus. Different FIA methods have been developed to solve analytical problems. In this work two operation modes have been tested: (i) normal FIA, where a carrier is used to transport the sample to the HPLC injector and (ii) completely continuous, where the sample solution is transported directly to the HPLC injector.[5]

The normal FIA method has been used to study the removal of pesticides from waste water by means of adsorption through activated carbon. A great variety of solutions containing phenols, chlorinated hydrocarbon and organophosphorus pesticides have been treated by carbon adsorption in the last years. This system appears to be the most effective and least expensive for removal of pesticides.[6-8]

In the present work, we try to confirm these observations and to prove that the FIA–HPLC is a suitable technique for the analysis of pesticides.

## EXPERIMENTAL

### Materials

Organophosphorus pesticides (azinphos methyl, diazinon and fenthion) were obtained from Bayer. Standard solutions were prepared by accurately weighing each pesticides into a 100 ml volumetric flask and diluting to volume with methanol.

Solvents were HPLC grade (Scharlau) and distilled water was microfiltered by Millipore filters (0.45 $\mu$m).

F-400 granular activated carbon (Panreac) with a particle size between 0.5 and 1 mm was used as the adsorbent in this study.

## Instrumentation

*High performance liquid chromatography*  The HPLC system consisted of a Spectra-Physics Model SP-8700 solvent delivery unit, an injection valve (Rheodyne) with a $10 \mu l$ sample loop, an UV–Vis detector (SP-8440), an integrator (SP-4270) and a Knauer Nucleosil-C18 column ($120 \times 4$ mm i.d.). The wavelength used was 220 nm. The mobile phase was 60:40 acetonitrile/water. The flow rate was set at 1 ml/min. Quantitative analyses were made using peak areas.

Under these conditions, the retention time of each pesticide was: diazinon (1.3 min), azinphos methyl (3.0 min) and fenthion (5.0 min).

*Flow injection analysis*  The FIA system consisted of a Watson-Marlow (202 U/AA4) peristaltic pump and an Omnifit $100 \mu l$ injection valve. Teflon tubing (0.5 mm i.d.) was used to connect the FIA to the HPLC. Silicone tubing (1.5–1.65 mm i.d.) was used as a column for the activated carbon.

## FIA techniques

*1) Completely continuous*  In this technique (Figure 1A) the main carrier is the standard solution. The injection valve of the FIA is not used. The solution is pumped and flows continuously through the injection valve of the HPLC apparatus and it can be injected at any time. The pesticide concentration can be determined without interruption.

*2) Normal FIA*  In this technique (Figure 1B) the carrier is distilled water and flows continuously through the HPLC apparatus. The standard solution is also pumped to the FIA injector. A well-defined volume of this standard solution is injected into the carrier stream at any time. The sample solution takes a certain time from FIA to HPLC injector. This time is a function of teflon tube length and FIA flow rate and should be perfectly determined in order to always inject the same amount of the sample through the HPLC column. Tube length has been kept constant ($122 \times 0.5$ mm i.d.) and the time has been calculated for several flow rates.

This operation mode can be used to analyze samples which have been treated before being injected into the carrier. One possibility is

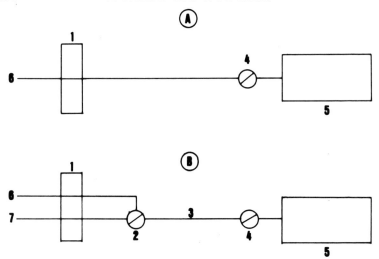

**Figure 1** Manifolds for the determination of pesticides. (A) continuous, (B) normal.
(1) peristaltic pump, (2) FIA injector, (3) teflon tube, (4) HPLC injector, (5) HPLC–
UV detector, (6) sample, (7) carrier, W: waste.

to recirculate this sample through an ion exchange or an adsorption
column. Figure 2 shows this recirculation mode. In this work, this
mode has been used to study the removal of pesticides from water
by adsorption on activated carbon. Measured amounts of the three
pesticides are poured into a flask containing distilled water. This
standard solution is pumped to the FIA injector in load position,
thus this solution circulates through the activated carbon column
and returns to the reservoir. The initial concentration in the reservoir was 10 mg/L for each pesticide. The relationship between mass
of activated carbon and sample volume was 10 mg/mL. The pesticide
concentration at any time can be obtained by injecting the sample
into the carrier and observing the HPLC chromatograms.

## RESULTS AND DISCUSSIONS

The detection limit was 0.5 mg/L in all cases.

Figure 3 shows some chromatograms of the same standard
solution obtained with the completely continuous mode. It can be

observed that the reproducibility of this mode is really good. Thus it can be useful for routine control of pesticides in industrial waste waters.

Figure 4 shows the chromatograms obtained by using the normal FIA mode for three different FIA flow rates. It can be observed that there are no response differences, so it is possible to select any of these conditions. In order to reduce consumption of standard solution the flow rate was restricted to 0.62 ml/min.

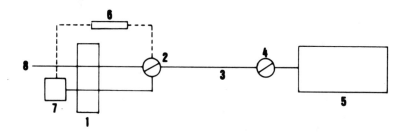

**Figure 2**   Manifold showing the recirculation mode. (1) peristaltic pump, (2) FIA injector, (3) teflon tube, (4) HPLC injector, (5) HPLC–UV detector, (6) Activated carbon column, (7) flask containing pesticide solution, (8) carrier, W: waste.

The chromatograms of both figures show narrow and symmetrical peaks. This means that there is no sample dispersion,[9] thus the FIA–HPLC technique allows to analyze these pesticides with the same sensibility as in conventional HPLC, but with an easier sample handling.

Figure 5 shows the decrease of pesticide concentration as a function of time. It can be observed that the adsorption percentage is different for each pesticide (diazinon 55%, azinphos methyl 60% and fenthion more than 90%). This can be related to the polarity of these compounds and thus their solubility in water (diazinon 40 mg/L, azinphos methyl 33 mg/L and fenthion 2 mg/L)[10] as there is a competition between adsorption and solubility, i.e. a higher solubility involves a lower adsorption.

Kobylinsky et al.[7] define the efficiency of the activated carbon column by means of an efficiency factor (k) by using a mass balance of the system.

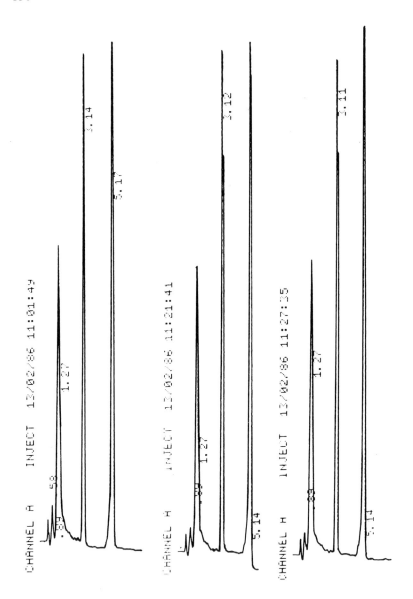

**Figure 3** Chromatograms obtained by using completely continuous FIA.

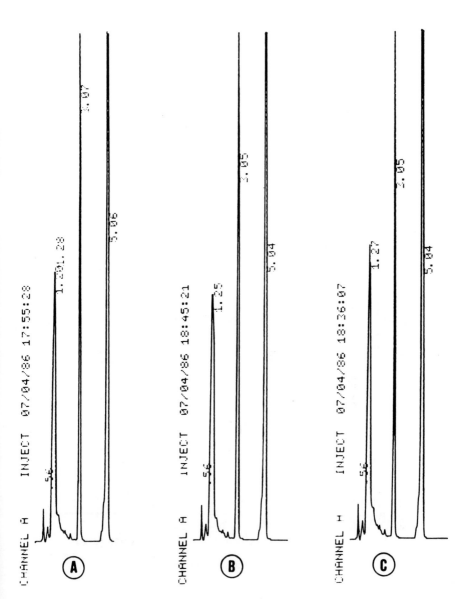

**Figure 4** Chromatograms obtained by using normal FIA at different flow rates. (A) 0.62, (B) 1.3 and (C) 2 ml min$^{-1}$.

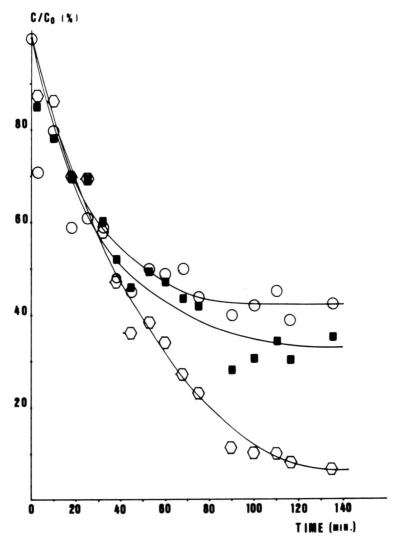

**Figure 5**   Concentration percentage *vs* time.
Diazinon ○
Azinphos-Me ■
Fenthion □

According to the following equation:

$$\log C/C_0 = -[kq/2.303\,V]t$$

Where $q$:  FIA flow rate (ml min$^{-1}$)
$\quad\quad V$:  sample volume (ml)
$\quad\quad C$:  concentration at any time (mg/L)
$\quad\quad C_0$: initial concentration (mg/L)
$\quad\quad t$:  time (min)

Plotting $\log C/C_0$ vs $t$, the efficiency factor can be obtained as the slope of the straight lines above the saturated time as shown in Figure 6. The $k$-values are: diazinon 0.24, azinphos methyl 0.20 and fenthion 0.35.

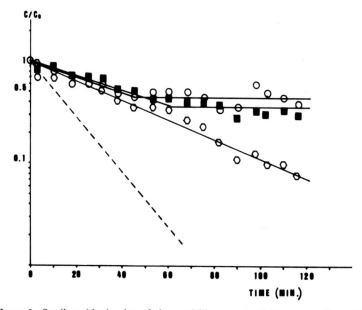

**Figure 6**  Semilogarithmic plot of the pesticide concentration *vs* time: Dashed line corresponds to $k = 1$.
Diazinon ○
Azinphos-Me ■
Fenthion

The efficiency factor can be improved by changing different parameters such as flow rate, relationship between volume and carbon mass, etc. These influences are being studied at the moment.

In this work, it has been proved that the combined FIA–HPLC is a simple and versatile technique because it facilitates the sample handling and can be used in routine analysis, as well as to study pesticide reactivity e.g. adsorption on activated carbon in this case.

## Acknowledgements

The authors are indebted to Professor Manuel Aguilar for his interest in this work.

## References

1. Ministerio de Industria y Energia (Spain), *Tecnologías Básicas Aplicables a la Depuración de los Efluentes Líquidos de la Industria de Plaguicidas* 1982.
2. J. Sherma and G. Zweig, *Anal. Chem.* **55,** 57R (1983).
3. J. F. Lawrence and D. Turton, *J. Chromatogr.* **159,** 207 (1978).
4. J. Ruzicka and E. H. Hansen, *Flow Injection Analysis* (J. Wiley and Sons, New York, 1981).
5. A. Rios, M. D. Luque de Castro and M. Valcarcel, *Talanta* **31**(9), 673 (1983).
6. *Treatment of Water by Granular Activated Carbon* (ACS series, 202, 1983).
7. M. Pirbazari and W. J. Weber Jr., *J. Environ. Eng.* **110**(3), 656 (1984).
8. E. A. Kobylinsky, W. H. Dennis Jr. and A. B. Rosencrace, *ACS series* **259,** 125 (1985).
9. M. Valcarcel and M. D. Luque de Castro, *Análisis por Inyección en Flujo* (Universidad de Córdoba 1984) p. 79.
10. C. R. Worthing, ed., *The Pesticide Manual* (British Crop Protection Council Publications, Worcestershire, 1983), 7th ed.

# Mass Spectrometric Determination of Tris(1,3-dichloro-2-propyl)-phosphate (TDCP) Using NCI-Technique[†]

ULLA SELLSTRÖM

*Department of Analytical Chemistry, University of Stockholm, S-106 91 Stockholm, Sweden*

and

BO JANSSON

*Special Analytical Laboratory, National Environmental Protection Board, Box 1302, S-171 25 Solna, Sweden*

(*Received August 17, 1986; in final form October 15, 1986*)

A number of different products suspected to contain flame retardants were analysed. TDCP was found in 11 out of 104 samples. It was most common in polyurethane products such as sound absorbing materials and liners for cars and buses. To get an integrated picture of TDCP exposure the contents of vacuum cleaner bags from two homes were analysed. One of these contained TDCP.

To investigate possible presence in humans, blood samples were analysed. For clean-up, Sep-Pak cartridges with $C_{18}$ phase were used. These cartridges contained TDCP and it was not possible to eliminate this background entirely.

The NCIMS detection sensitivity for TDCP was better when analysing plasma extracts than when analysing pure standard solutions. Thus quantitative determinations of TDCP must be made by standard additions to plasma extracts.

Something in the blood plasma seems to be able to bind or immobilise a certain

---

†Presented at the Workshop on Chemistry and Fate of Organophosphorus compounds, Amsterdam, June 18–20, 1986.

This article was first published in *International Journal of Environmental Analytical Chemistry*, Volume 29, Number 4 (1987).

amount of TDCP. Therefore it was not possible to analyse amounts less than 600 pg/ml whole blood with this method. None of the 37 analysed blood samples exceeded this value.

KEY WORDS:     Tris(1,3-dichloro-2-propyl)phosphate, flame retardant, analysis, mass spectrometry, NCIMS.

## INTRODUCTION

TDCP is used as a flame retardant mainly in polyurethane foam and polyester textiles. Animal experiments have shown that TDCP is readily absorbed from the skin and gastrointestinal tract of rats and rapidly distributed throughout the body. It is subject to rapid and extensive metabolic degradation. Less than 19% of the administered TDCP could be recovered as TDCP 30 min after intra-venous dosing.[1,2] TDCP and some of its metabolites have been reported to be mutagenic.[1-6]

The water solubility of TDCP is approximately 100 ppm, and it is relatively stable in water after 24 hours.[7,8] It is toxic to killifish and goldfish, the $LC_{50,96h}$ values being 3.6 ppm and 5.1 ppm, respectively.[9] In another investigation, 6 out of 6 goldfish were found dead after 24 hours with a concentration of 5 ppm TDCP in the water.[8] In the same work, $LD_{50}$ for rats was found to be 2830 mg TDCP/kg.

TDCP has been found in Canadian drinking water[10,11] and in Japanese water and sediment samples.[12,13] It has also been found in Canadian human adipose tissues. In 5 out of 16 samples the amount of TDCP was between 0.5 to 110 ng/g.[14] In the USA, TDCP was found in human seminal plasma. It was found in 34 of the 123 analysed samples, in the range of 5 to 50 ppb.[15]

This work was performed in order to investigate the present use of TDCP in Sweden and its presence in humans. Mass spectrometry of negative ions from chemical ionisation (NCIMS) shows a good sensitivity for the detection of TDCP. Detection limit was 10 pg.

## EXPERIMENTAL

### Gas Chromatography/Mass Spectrometry (GC/MS)

The GC/MS system consisted of a Finnigan Model 9610 gas

chromatograph interfaced to a Finnigan Model 4021 quadrupole mass spectrometer, equipped with a 4500 ion source (exchangeable ion volumes). It was operated in the chemical ionisation (CI) mode. Data were acquired using an Incos Data System.

The GC was equipped with a DB-5 (0.25 μm) fused silica capillary column (25 m × 0.32 mm i.d.) directly fitted into the CI source. Transfer line temperature was 200°C and ion source temperature was 80°C.

Vaporisation injection in the splitless mode was used for sample introduction. Injector temperature was 210°C. Injections were made at a column temperature of 80°C when keeper was added, otherwise at 60°C. This temperature was maintained for 1 min and was followed by a 10°C/min linear temperature program to a maximum of 280°C, which was held for 10 min.

High purity helium (AGA, Stockholm) was used as carrier gas. Methane (99.95%, AGA, Stockholm) was used as reagent gas at a pressure of 0.40 torr. Emission current was 0.2 mA, electron multiplier voltage was 1300 V, and electron energy was 70 eV.

The only ion of importance for TDCP was the $(M-111)^-$ ion, corresponding to loss of a dichloropropyl group from the molecular ion. Measured ions in the mass fragmentogram analyses are shown in Table I. The ions m/z $319^-$, $263^-$, and $292^-$ were used for the quantitative analysis.

## Gas Chromatography/Electron Capture Detection (GC/ECD)

A Varian Model 3700 gas chromatograph equipped with a $^{63}$Ni detector and a DB-5 (0.25 μm) fused silica capillary column

**Table I**  Ions used for mass fragmentographic analysis of TDCP

| Compound | m/z |
|---|---|
| TDCP | $317^-$, $319^-$, $321^-$, $323^-$ |
| Methylparathion (Internal standard) | $141^-$, $154^-$, $263^-$ |
| 2,2',5,6',-tetrachlorobiphenyl (Injection standard) | $290^-$, $292^-$, $294^-$, $296^-$ |

F

(5 m × 0.32 mm i.d.) was used. Splitless injections were made at an injector temperature of 210°C and a column temperature of 60°C. The column temperature was maintained for 1 minute, followed by a 6°C/min linear temperature program to a maximum of 280°C, which was held for 5 min. Detector temperature was 300°C. High purity nitrogen (AGA, Stockholm) was used as carrier gas.

## Materials

TDCP (Franzén & Fried). Methylparathion (US EPA). Methanol (Merck, p.a.). Sep-Pak $C_{18}$ (Waters Associated Millipore Ltd.). 2,2',5,6'-tetrachlorobiphenyl (CA Wachtmeister, University of Stockholm). Bloodtainer, vacuum blood collecting system, 10 ml tubes (testab laboratorieprodukter AB, Huskvarna, Sweden). n-Hexane and n-undecane were redistilled in our laboratory.

## Products

Most products were bought in shops and supermarkets in or near Stockholm. Products suspected for containing flame retardants were chosen (Table II).

The contents of vacuum cleaner bags from two homes were analysed in order to get an integrated picture of TDCP exposure. One house was recently built and the other one was about 15 years old.

## Blood samples

Most blood donors were employees at the National Environmental Protection Board working in the office, but some worked in our laboratory where TDCP containing materials are used for sound absorbing.

Blood samples (20 ml) in two bloodtainer tubes from each of totally 37 donors were collected. The samples were centrifuged the same day and the plasma was transferred to sample tubes with teflon-lined screwcaps. Samples that were subject to clean-up the following day were placed in a refrigerator overnight, whereas storage for longer periods was done in a freezer. All samples were cleaned-up within one week. Methylparathion was added immediately before clean-up.

**Table II** Type of samples analysed for content of TDCP

| Classes of products | Number | Positive response | |
| --- | --- | --- | --- |
| | | GC/ECD | GC/NCIMS |
| Sound absorbing materials | 7 | 6 | 6 |
| Shock absorbing materials | 8 | 1 | 1 |
| Mattresses | 12 | 1 | 1 |
| Wall papers | 3 | – | – |
| Curtains | 1 | – | – |
| Bedding materials | 5 | – | – |
| Carpets | 4 | – | – |
| Furnishing fabrics | 21 | – | – |
| Bus liners | 11 | 2 | 2 |
| Car liners | 6 | 1 | 1 |
| Aircraft fabrics | 1 | – | – |
| Childrens clothes | 4 | – | – |
| Childrens toys | 6 | – | – |
| Ironing board covers | 1 | – | – |
| Lamp-shades | 1 | – | – |
| Kettle-holders | 2 | – | – |
| Diffusor filters (vacuum cleaners) | 1 | – | – |
| Working clothes | 3 | – | – |
| Buildings materials | 5 | – | – |
| Ear plugs | 1 | – | – |
| Commercial flame retardants | 1 | – | – |
| | 104 | 11 | 11 |

## Analyses of products

The products were extracted with $n$-hexane and analysed qualitatively by GC/ECD. Confirmation of the presence of TDCP was made by GC/NCIMS.

## Analyses of blood samples

A scheme of the method is shown in Figure 1. The blood (20 ml) was centrifuged and 1.14 ng of methylparathion (internal standard) in $100\,\mu$l of $n$-hexane was added to the plasma. The hexane was evaporated with nitrogen. For the clean-up procedure Sep-Pak $C_{18}$

A N A L Y S I S   O F   B L O O D   S A M P L E S

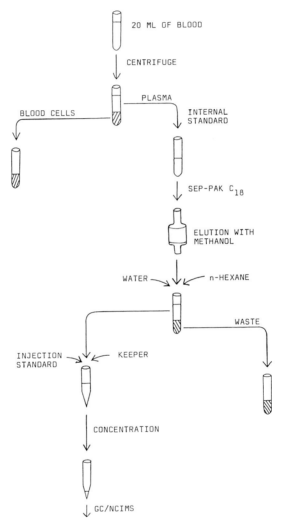

**Figure 1**   Schematic of the clean-up method for blood samples.

cartridges were used. These were Soxhlet extracted for 24 hours with methanol. Immediately before use they were washed with 8 ml of methanol, followed by 5 ml of water. Then the plasma was slowly pumped through the cartridge with a syringe. Following this, 3 ml of water was pumped through the cartridge which was then eluted with 2.5 ml of methanol. To the methanol 0.5 ml of water and $2 \times 1.5$ ml of n-hexane were added. To the n-hexane phase $25 \mu l$ of n-undecane (keeper) and $1.25 \mu g$ of 2,2′,5,6′-tetrachlorobiphenyl (injection standard) in $100 \mu l$ of n-hexane were added. The volume was reduced to approximately $25 \mu l$ by blowing with nitrogen and then analysed by GC/NCIMS.

Blanks were made in the same way using 20 ml of water which had been allowed to stand for a few days in the same type of bloodtainer tubes as were used for blood sampling. The methylparathion, however, was added after the clean-up.

The clean-up efficiency was evaluated by spiking a number of plasma samples (from 20 ml blood) from the same person with 500 pg, 1 ng, 2 ng and 5 ng TDCP, respectively, before clean-up. For comparison, other plasma samples were spiked with the same amounts after clean-up. One sample in each series was left unspiked.

## RESULTS AND DISCUSSION

Due to the presence of TDCP containing materials in the laboratory, the air was checked as follows. An open vessel with 100 ml of iso-octane was allowed to stand in the laboratory for about 40 hours. The volume was reduced to approximately 1 ml by blowing with nitrogen and then $2 \mu l$ was injected on the GC/ECD. No TDCP was found.

### Product analysis

Qualitative product analyses were made after extracting the different samples with n-hexane. The extracts were then injected on GC/ECD and confirmation of the presence of TDCP was made by GC/NCIMS (Table II). TDCP was found only in different kinds of polyurethane foam, such as sound absorbing materials and liners for cars and buses. In Figure 2 are shown chromatograms and mass

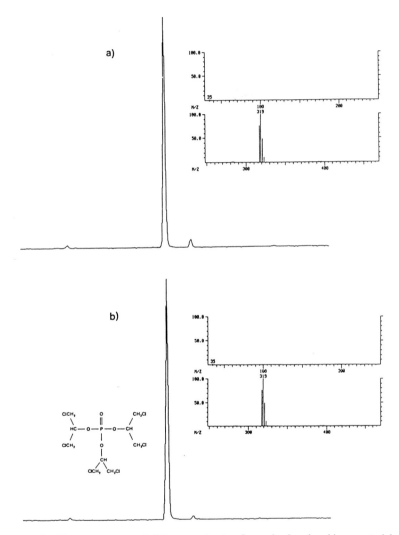

**Figure 2** Chromatograms of (a) an extract of a shock absorbing material (polyurethane) and (b) a TDCP standard, and mass spectra of the major peaks. Chromatographic and mass spectrometric conditions are given in the text.

spectra for a polyurethane sample and a TDCP standard, respectively.

The content of one of the vacuum cleaner bags (from the older house) contained TDCP, while the other one did not.

## Blood analysis

The conditions for the chromatography are influenced in a positive way by components in the plasma. The NCIMS detection sensitivity for TDCP and methylparathion was also much better when analysing blood extracts than when analysing pure standard solutions. Because of this effect, quantitations must be made by standard additions of TDCP to plasma extracts.

The plasma samples spiked with 500 pg, 1 ng and 2 ng TDCP, respectively, gave approximately the same response relative to the injection standard as the unspiked sample. The sample spiked with 5 ng TDCP gave a higher value. The samples spiked after the Sep-Pak clean-up gave a fairly good calibration curve. These results indicate that there is something in the blood plasma that is able to bind or immobilise a certain amount of TDCP.

In Figure 3 mass fragmentograms of a blood sample spiked with TDCP, an unspiked blood sample and a blank, are shown. The Sep-Pak cartridges contained TDCP and it was not possible to eliminate this background entirely although they were Soxhlet extracted with methanol for 24 hours. Washing with ethanol and Soxhlet extraction with n-hexane did not improve the blanks. After Soxhlet extracting with methanol for 24 hours the background was, however, at a relatively constant level.

Due to the background problems and bad recoveries of spiked TDCP only levels above 600 pg/ml whole blood could be analysed. None of the 37 analysed blood samples exceeded this value.

## CONCLUSIONS

TDCP does not seem to be commonly used in Sweden. Its usage seems to be limited to specific product groups.

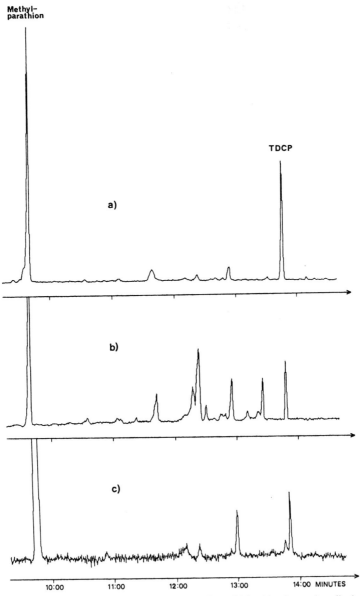

**Figure 3** Mass fragmentograms (m/z $319^- + 263^-$) of (a) a blood sample spiked with TDCP; (b) an unspiked blood sample; and (c) a blank. Chromatographic conditions are given in the text.

TDCP and methylparathion both chromatograph better in blood extracts than in pure solvents. The detection sensitivity using NCIMS is also better in blood extracts.

TDCP in blood plasma is lost to a certain degree. It is therefore not possible to detect levels of TDCP below 600 pg/ml whole blood with this method.

## Acknowledgement

We are indebted to Lars Ehrenberg and Tomas Alsberg for valuable discussions. This investigation was performed with support from the Products Control Board in Sweden.

## References

1. A. A. Nomeir, S. Kato and H. B. Matthews, *Toxicol. Appl. Pharmacol.* **57,** 401 (1981).
2. R. K. Lynn, K. Wong, C. Garvie-Gould and J. M. Kennish, *Drug Metab. Disp.* **9,** 434 (1981).
3. E. J. Söderlund, E. Dybing, J. A. Holme, J. K. Hongslo, E. Rivedal, T. Sanner and S. D. Nelson, *Acta Pharmacol. et Toxicol.* **56,** 20 (1985).
4. A. Nakamura, N. Tateno, S. Kojima, M.-A. Kaniwa and T. Kawamura, *Mutation Res.* **66,** 373 (1979).
5. D. Brusick, D. Matheson, D. R. Jagannath, S. Goode, H. Lebowitz, M. Reed, G. Roy and S. Benson, *J. Environm. Path. Toxicol.* **3,** 207 (1979).
6. M. D. Gold, A. Blum and B. N. Ames, *Science* **200,** 785 (1978).
7. V. D. Ahrens, G. A. Maylin, J. D. Henion, L. E. St. John, Jr and D. J. Lisk, *Bull. Environm. Contam. Toxicol.* **21,** 409 (1979).
8. A. T. Eldefrawi, N. A. Mansour, L. B. Brattsten, V. D. Ahrens and D. J. Lisk, *Bull. Environm. Contam. Toxicol.* **17,** 720 (1977).
9. K. Sasaki, M. Takeda and M. Uchiyama, *Bull. Environm. Contam. Toxicol.* **27,** 775 (1981).
10. D. T. Willams and G. L. LeBel, *Bull. Environm. Contam. Toxicol.* **27,** 450 (1981).
11. G. L. LeBel, D. T. Williams and F. M. Benoit, *J. Assoc. Off. Anal. Chem.* **64,** 991 (1981).
12. S. Ishikawa, T. Taketomi and R. Shinohara, *Water Res.* **19,** 119 (1985).
13. S. Ishikawa, pers. communication, 1986.
14. G. L. LeBel and D. T. Williams, *J. Assoc. Off. Anal. Chem.* **66,** 691 (1983).
15. T. Hudec, J. Thean, D. Kuehl and R. C. Dougherty, *Science* **211,** 951 (1981).

# The Application of a Flame Photometric Detector in Packed Microcapillary Liquid Chromatography: Detection of Organophosphates†

CHARLES E. KIENTZ and ALBERT VERWEIJ

*Prins Maurits Laboratory TNO, P.O. Box 45, 2280 AA Rijswijk, The Netherlands*

(*Received November 19, 1986; in final form January 12, 1987*)

A commercially available flame photometric detector (FPD) designed for gas chromatography has been investigated for its applicability in microcolumn liquid chromatography. The column effluent is evaporated and subsequently introduced into the detector using a home-made interface. The influence of the mode of effluent introduction into the flame, the composition and flow rates of the flame gases and eluent on the detector performance are discussed.

KEY WORDS:   Micro HPLC, organophosphates, flame photometric detector (FPD).

## INTRODUCTION

The development of miniaturized HPLC has allowed the direct introduction of the entire liquid effluent into gas chromatographic detectors, which has led to the successful implementation of phosphorus-sensitive detectors in small bore HPLC. Among these detectors, the thermionic detector (TID) recently received considerable attention. Basic research in this field was done[1-4] using packed microcapillary columns coupled via a nebulization interface introducing the effluent directly into a dual-flame TID. A different

---

†Presented at the Workshop on Chemistry and Fate of Organophosphorus Compounds, Amsterdam, June 18–20, 1986.

This article was first published in *International Journal of Environmental Analytical Chemistry*, Volume 30, Numbers 1 + 2 (1987).

approach was made[5-7] using narrow-bore columns and an interface to vaporize the LC effluent before entering the single flame TID.

However, the use of a flame photometric detector (FPD) in microcolumn liquid chromatography has been described only in few papers. A miniaturized FPD was described by McGuffin *et al*[8] who was able to detect 2 ng of phosphorus by means of direct introduction of the column effluent into the flame. At the tenth International Symposium on Column Liquid Chromatography in San Francisco (1986) the Varian Instrumentation Group presented a dual flame micronebulizer-FPD as an HPLC detector.[9] In our laboratory we are dealing with the analysis of polar, acidic and other high-boiling phosphorus containing compounds. Generally, HPLC-UV detection and GC analysis are not suitable for such compounds. Therefore we are investigating the possibility of coupling packed capillary fused silica columns to a commercially available FPD detector to determine phosphorus directly in the HPLC effluent. Some preliminary results are given in this paper.

## EXPERIMENTAL

### Instrumentation

The system was assembled from a Shimadzu LC-5A pump, a Valco sample valve with 60 nl internal volume and a Tracor FPD 100 AT. The fused silica capillaries were packed with Lichrosorb RP 18 particles (10 $\mu$m) according to the procedure of Gluckman *et al*.[10] The inlet side of the capillary column was directly connected to the injector using a finger-tightened PTFE ferrule and nut (Hibar, Merck, F.R.G.). The column performance was tested separately using a home-made 40 nl micro flow UV-cell.[11]

### Materials

The solvents were of HPLC-grade quality supplied by Merck (Darmstadt, F.R.G.). Trimethyl phosphate (TMP, b.p. 197°C; Merck, Darmstadt, F.R.G.), triethyl phosphate (TEP, b.p. 215°C; BDH Ltd, Poole, England), tributyl phosphate (TBP, b.p. 289°C; UCB, Brussels, Belgium) and triphenyl phosphate (TPP, b.p. 245°C 11 mm Hg; Aldrich Europe, Beerse, Belgium) were of analytical grade. The adhesives used were epoxy glue (UHU, Linger + Fischer GmbH) and Silcoset 151 (Ambersil Ltd, England).

## Construction of the interface

The design of the interface between the LC fused silica capillary column and the FPD detector is shown in Figure 1. The packing at the end of the column was confined by a porous Teflon frit 0.2 mm secured by two internally close-fitting fused silica capillaries.[11]

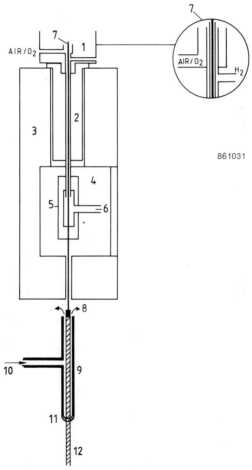

**Figure 1** Column-detector interface. 1 = burner base, 2 = aluminium cylinder (length 80 mm), 3 = detector oven body, 4 = detector oven, 5 = T-piece, 6 = nitrogen purge, 7 = tip of flame base with fused silica capillaries (0.32 mm I.D., 0.1 mm I.D.), 8 = end of 12, 9 = cooling device (GC-splitter), 10 = air cooling, 11 = Silcoset adhesive, 12 = packed capillary column.

The column [12] is fixed with Silcoset 151 adhesive [11] in a cooling device derived from an original GC-splitter. The purpose of the cooling device is to prevent too premature an evaporation of the effluent outside the hot detector oven body [3] due to heat radiation. The column end [8] is connected to a fused silica capillary (I.D. 0.1 mm) which is inserted through the detector oven [4] and the burner base [1] directly into the flame jet. Besides a nitrogen purge [6] flowing through a fused silica capillary of 0.32 mm I.D. [7] is added by means of a T-piece [5].

## RESULTS AND DISCUSSION

The performance of the micro HPLC-FPD system was investigated by varying parameters such as the adjustment of the effluent into the flame jet, the flame gas and LC eluent flow rates, and the eluent composition. A characteristic chromatogram of a mixture of four organophosphates is shown in Figure 2.

The observed reduced plate height ($h$), peak asymmetry ($A$) and detector sensitivity $(Sm)^{12}$ of the phosphorus compounds are listed in Table I. Under identical LC conditions using a 40 nl microflow UV cell at 260 nm[11] the reduced plate height of TPP was 7.9. This value corresponds with those of TMP and TEP given in Table I. This may indicate that on application of the micro LC-FPD equipment the increased peak-broadening, peak asymmetry and the detector sensitivity for TBP and TPP are probably due to insufficient evaporation of these relatively high-boiling compounds in the interface used.

**Table I**  Micro LC results as presented in Figure 2

| Compound | $h$ | $A$ | $Sm(\text{ng } P/s)$ |
|----------|------|-----|------------|
| 1 TMP | 7.4 | 1.0 | 5.7 |
| 2 TEP | 7.0 | 1.0 | 6.4 |
| 3 TPP | 15.2 | 1.5 | 9.5 |
| 4 TBP | 15.7 | 3.8 | 7.8 |

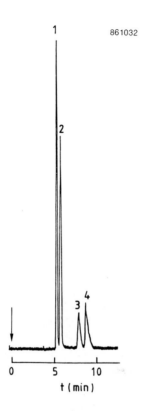

**Figure 2** LC-FPD Chromatogram of a mixture of four organophosphates column 500 mm × 0.5 mm × 0.32 mm (L × O.D. × I.D.) packed with Lichrosorb RP 18, 10 μm.; mobile phase acetonitrile/water 9:1; flow 5 μl/min; interface temp. 310°C; air flow rate 250 ml/min; hydrogen flow rate 360 ml/min; oxygen flow rate 5 ml/min.; compounds: 1 = TMP, 2 = TEP, 3 = TPP, 4 = TBP.

## Adjustment of the fused silica capillary into the flame jet

The adjustment of the effluent outlet (Figure 1, item 7) near the flame jet appeared to be very critical as can be seen from Figure 3. Placing the end of the capillary 40–50 mm below the flame tip will be a compromise in the optimization of the different responses.

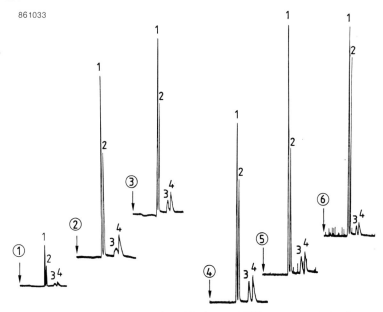

**Figure 3** Influence of column-end positioning on FPD responses. 1 = in the top of the flame jet, 2 = 20 mm, 3 = 30 mm, 4 = 40 mm, 5 = 50 mm, 6 = 60 mm below the flame jet. See for other conditions Figure 2.

## Flame gas flow rates

The performance of the FPD was evaluated varying the hydrogen, air and oxygen flow rates. The influence of different hydrogen flow rates on peak shape and sensitivity of the organophosphates was insignificant. A stable flame was obtained over 170 ml/min, below 170 the flame extinguished. The air flow rate proved to be an important parameter influencing the peak height as presented in Figure 4.

From experiments represented in Figure 4 it can be seen that when the air flow increased the response of TMP (peak 1) improved a factor of 30. When the oxygen flow was subsequently increased to a total hydrogen-to-oxygen ratio of 3.6 the peak height improved additionally a factor of 10. This hydrogen-to-oxygen ratio is in accordance with the optimum value given by McGuffin et al.,[8] which is comparable with the value when using this detection system in gas chromatography.

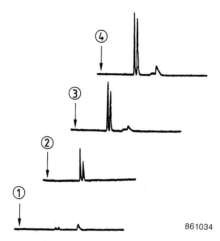

861034

**Figure 4** Influence of air flow rate on detector response. 1 = 110 ml/min, 2 = 140 ml/min, 3 = 175 ml/min, 4 = 215 ml/min (H2/02 ratio is 8.4); Hydrogen flow = 360 ml/min; other conditions see Figure 2.

## Eluent composition

Two eluent compositions were compared: acetonitrile-water and methanol-water, ratio 8:2. Comparable signal-to-noise ratios were obtained on analysing TMP and TEP using both eluent compositions as can be seen from Table II.

**Table II** Influence of eluent composition on signal-to-noise ratio

| Compound | Eluent composition | |
|---|---|---|
| | Acetonitrile/water 8:2 | Methanol/water 8:2 |
| | Signal/noise | Signal/noise |
| TMP | 11 | 13 |
| TEP | 3 | 3 |

In both cases a similar noise level of 1.5% full scale (att. $10^4 \times 2$, recorder 1 mv) was measured at flow rates of 20 μl/min. Injected amounts 14 μg TMP and 6 μg TEP.

The result from Table II differs from those obtained by McGuffin[8] and Chester[13,14] who found increased background noise substituting methanol for acetonitrile even at very low 1–5% concentrations in aqueous mixtures.

## Eluent flow rate

From the detector sensitivity values for TEP given in Table III it can be concluded that the FPD response depends on the eluent flow rate. Additionally, it was found that by decreasing the flow rate an improvement in response could be obtained both for TMP and TEP, whereas TPP and TBP gave poorly shaped peaks. The strong dependence of the FPD response on the flow rate indicates that especially for quantitative purposes a very stable eluent flow will be necessary.

**Table III** Influence of eluent (acetonitrile: water 8:2) flow rate on detector[a] sensitivity (Sm) of TEP

| Flow rate ($\mu$l/min) | Sm(ng P/s) |
|---|---|
| 1 | 0.1 |
| 2 | 1.6 |
| 20 | 110.0 |

[a]H2/O2 ratio = 3.6.

## Nitrogen purge

Addition of nitrogen purge flow showed an improved peak height and shape of the phosphorus containing compounds.

## EVALUATION

In view of the above-mentioned preliminary and orientating experiments with the micro LC-FPD equipment further investigations will be concentrated on improving the interface to obtain sufficient evaporation of high-boiling phosphorus-containing compounds and subsequently, efficient introduction into the flame. The distance between column and flame tip, the hydrogen-to-oxygen ratio, the eluent flow rate and nitrogen purge flow rate are important parameters to be studied more thoroughly.

## Acknowledgement

We are indebted to Jan de Bruin for his experimental assistance.

## References

1. V. L. McGuffin and M. V. Novotny, *J. Chromatogr.* **218,** 179 (1981).
2. V. L. McGuffin and M. V. Novotny, *Anal. Chem.* **55,** 2296 (1983).
3. J. C. Gluckman and M. V. Novotny, *J. Chromatogr.* **314,** 103 (1984).
4. J. C. Gluckman and M. V. Novotny, *J. Chromatogr.* **333,** 291 (1985).
5. F. A. Maris, R. J. van Delft, R. W. Frei, R. B. Geerdink and U. A. Th. Brinkman, *Anal. Chem.* **58,** 1634 (1986).
6. J. C. Gluckman, D. Barceló, G. J. de Jong, R. W. Frei, F. A. Maris and U. A. Th. Brinkman, *J. Chromatogr.* (1986) in press.
7. D. Barceló *et al., Int. J. of Env. Anal. Chem.* (1986) in press.
8. V. L. McGuffin and M. Novotny, *Anal. Chem.* **53,** 946 (1981).
9. J. F. Karnicky and S. van der Wal, presented at the tenth International Symposium on Column Liquid Chromatography, San Francisco (1986).
10. J. C. Gluckman, A. Hirose, V. L. McGuffin, M. Novotny, *Chromatographia.* **17,** 303 (1984).
11. C. E. Kientz and A. Verweij, to be published.
12. R. P. W. Scott, *Liquid Chromatogr. Detectors* (Elsevier Sc. Publ. Co., Amsterdam, 1977) Chap. 2, p. 18.
13. T. L. Chester, *Anal. Chem.* **52,** 638 (1980).
14. T. L. Chester, *Anal. Chem.* **52,** 1621 (1980).

# Determination of Trialkyl and Triaryl Phosphates by Narrow-bore Liquid Chromatography with On-line Thermionic Detection†

## D. BARCELÓ, F. A. MARIS‡, R. W. FREI, G. J. de JONG and U. A. Th. BRINKMAN

*Section of Environmental Chemistry, Department of Analytical Chemistry, Free University, De Boelelaan 1083, 1081 HV Amsterdam, The Netherlands*

(*Received December 15, 1986*)

The potential of a previously reported on-line reversed-phase liquid chromatography-thermionic detector (LC-TID) system has been further evaluated. Several trialkyl (trimethyl to tri-octyl) and triaryl (triphenyl and tri-o-cresyl) phosphates were chosen as model compounds. LC was done on an alkyl-bonded silica column with methanol-water (80:20) at a flow-rate of 35–70 $\mu$l/min as eluent. The band broadening in the interface-TID unit increased with decreasing volatility of the compounds and for the relatively non-volatile compounds, the band broadening decreased with increasing eluent flow-rate.

For most of the test compounds, detection limits of 40–100 pg were obtained. As an illustration of the high selectivity of the LC-TID system, the trace-level determination of two aryl phosphates in sediment samples is reported.

---

†Presented at the Workshop on Chemistry and Fate of Organophosphorus Compounds, Amsterdam, Holland, June 18–20, 1986.

‡Author to whom all correspondence should be addressed.

This article was first published in *International Journal of Environmental Analytical Chemistry*, Volume 30, Numbers 1 + 2 (1987).

## INTRODUCTION

Miniaturization in column liquid chromatography (LC) has eliminated many of the difficulties associated with direct introduction of LC eluents into gas chromatographic detectors. The utilization of detectors such as the electron-capture detector,[1,2] the mass spectrometer,[3,4] the flame emission[5] and thermionic detector[6] for narrow-bore LC has already been described. We recently published[7,8] our first results about the coupling of a narrow-bore LC system with a thermionic detector (TID). The possibility of handling mobile phase flow-rates of up to $40 \mu l/min$ of methanol-water was shown, and the organophosphorus pesticides coumaphos, azinphos-ethyl and trichlorfon were determined in cabbage and tomato samples. Detection limits of 30 to 100 pg of injected compound or 0.2–0.5 pg/s of phosphorus were obtained.

In the present study, a series of thermally stable trialkyl and triaryl phosphates was selected as model compounds, to test the applicability of the LC-TID interface performance for compounds of decreasing volatility.

Most of the selected phosphates are of environmental interest. Major uses of trialkyl and triaryl phosphates are as fire retardant plasticizers in the manufacture of PVC plastics and as industrial hydraulic fluids to replace polychlorinated biphenyls. The estimated production of the phosphates is approximately $77 \times 10^6$ kg for 1986. This large volume results in substantial release to the environment either from losses during manufacturing or use.[9,10] Residues of triphenyl phosphate (TPhP) and tri-o-cresyl phosphate (TOCP) have been found in several environmental samples such as fish, sediment and water.[11-13] Recently, the highly toxic TOCP was also found in contaminated edible oils.[14] In the present paper, the determination of TPhP and TOCP in sediment by LC-TID is reported.

## EXPERIMENTAL

### Materials

All solvents were of HPLC-grade quality (Baker, Deventer, The Netherlands) and were passed through a filter of $0.45 \mu m$ pore diameter before use. Analytical reagent grade trimethyl phosphate

(TMP), triethyl phosphate (TEP), tripropyl phosphate (TPP), tributyl phosphate (TBP), tri-octyl phosphate (TOP), triphenyl phosphate and tri-o-cresyl phosphate were obtained as gifts from A. Verweij (TNO, Rijswijk, The Netherlands).

## Sample preparation

Sample pre-treatment was carried out using a simple procedure described in the literature[11] for the extraction of triaryl phosphates from sediment samples. The procedure is as follows: 25 g of sediment (wet weight) spiked with 2 ppm of TPhP and TOCP were placed in a soxhlet and 150 ml of methanol-water (9:1) were used as extraction solvent for a period of 4 hours. Afterwards, the aqueous methanol was evaporated in a rotary evaporator to a volume of 2 ml.

## Chromatographic system

Glass-lined stainless-steel columns (GLT; 200 × 0.7 or 1 mm I.D., SGE, Melbourne, Australia), laboratory packed with 5 $\mu$m Spherisorb ODS 2 (Merck, Darmstadt, F.R.G.) were used. Band-broadening studies were carried out using a flow injection system with the injection valve directly coupled to the evaporation interface. A Gilson (Villiers-le-Bel, France) Model 302 high-pressure pump coupled with a laboratory-made membrane pulse damper, or a $\mu$LC-500 high pressure syringe pump from ISCO (Lincoln, NE, U.S.A.) provided stable eluent delivery. Samples were introduced via a laboratory-made injection valve having a 0.05 and a 0.5 $\mu$l internal loop or a Valco (Houston, TX, U.S.A.) C214W micro valve with a 0.1 $\mu$l internal volume rotor. In flow injection analysis, the test compounds were dissolved in the mobile phase.

## Thermionic detector

A Packard (United Technologies, Delft, The Netherlands) Model 427 gas chromatograph was used, which contained a Packard Model 905 TID, the rubidium bead of which was electrically heated by a Packard Model 612 detector controller. The evaporation interface was located in the detector block, which was maintained at 300°C. The termination of the LC column was directly connected to the

D. BARCELÓ *ET AL.*

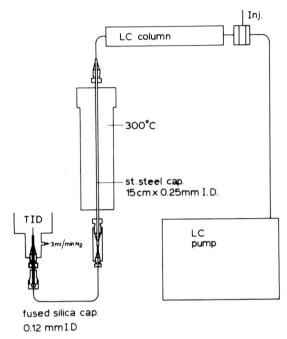

**Figure 1** Schematic design of the LC-TID system, also showing the detailed construction of the interface.

15 cm × 0.25 mm I.D. stainless-steel interface and the vaporized eluent from the interface was introduced into the flame jet via a 15 cm × 0.12 mm I.D. fused silica capillary; a nitrogen flow of 3 ml/min was added via a T-piece just below the detector body. This gas flow was pre-heated in the GC oven, in order to minimize its cooling effect. A schematic diagram of the LC-TID system used in the present study with the detailed construction of the interface is shown in Figure 1 (see also Reference 8).

## RESULTS AND DISCUSSION

### Band broadening in the interface

An important criterion for the performance of the LC-TID interface is the contribution to the band broadening of the analytes. A flow injection system is suitable to measure the band broadening in the interface and detector. It is well known that with this method tailing peaks are often obtained; therefore, band broadening was calculated by means of the Foley–Dorsey[15] experimental approximation of the second moment $(M_2)$.

In a previous study,[8] $M_2$ was found to be $0.66\,\mu l^2$ and $1.78\,\mu l^2$ at a flow-rate of $20\,\mu l/min$ for paraoxon-ethyl and coumaphos, respectively. These numbers demonstrate that the band broadening increases with decreasing volatility of the analytes. Table I, in which $M_2$ is expressed in both $\mu l^2$ and $s^2$, shows the same phenomenon for the trialkyl and triaryl phosphates studied in this paper. Table I also shows that an increase of the flow-rate causes an increase of $M_2(\mu l^2)$ for the relatively volatile test compounds, while a decrease of $M_2$ is observed for the less volatile compounds such as coumaphos, TPhP, TOCP and TOP. The band broadening $M_2(s^2)$ decreases for all compounds at higher flow-rates because of the decrease of the residence time in the interface and the detector.

A probable explanation for the influence of the flow-rate on the band broadening $(M_2; \mu l^2)$ of the relatively non-volatile test solutes is via the vaporization efficiency of the interface. Generally speaking,

**Table I**  Dependence of second moments $(M_2)$ of several phosphates on mobile-phase flow-rate. Conditions: Carrier stream, methanol-water (80:20). Interface temperature, 300°C. Experiments: $n=4$, rel. S.D. $=3.5$–$4.5\%$. $M_2$ measured for TID interface plus detector

| Compound | Boiling point (°C) | $M_2(s^2)$ at flow-rate ($\mu l/min$) | | | | $M_2(\mu l^2)$ at flow-rate ($\mu l/min$) | | | |
|---|---|---|---|---|---|---|---|---|---|
| | | 20 | 30 | 40 | 60 | 20 | 30 | 40 | 60 |
| Paraoxon-ethyl | | 5.9 | 3.5[a] | | | 0.66 | 1.19[a] | | |
| Coumaphos | | 16.1 | 4.1[a] | | | 1.78 | 1.38[a] | | |
| TMP | 197 | 5.3 | | 2.1 | | 0.58 | | 0.93 | |
| TEP | 216 | 5.4 | | 2.2 | | 0.60 | | 0.97 | |
| TBP | 289 | 5.6 | | 2.4 | | 0.62 | | 1.06 | |
| TPhP | 390 | | 39 | 17 | 6.5 | | 9.8 | 7.4 | 6.5 |
| TOCP | 410 | | 40 | 18 | 6.5 | | 10 | 7.8 | 6.5 |
| TOP | 216 (5 mm Hg) | | 101 | 52 | 21 | | 25 | 23 | 21 |

[a]Flow-rate, 35 $\mu l/min$.

an increase of the flow-rate may be expected to detract from an efficient evaporation. However, heat from the interface can easily radiate towards the separation column, creating a temperature gradient in the connecting capillary. Therefore, compounds leaving the column experience a gradual warming up, instead of a sharp temperature increase, and are vaporized over a broad zone. This problem will be especially severe with high-boiling substances, which may be incompletely vaporized in the low-temperature region of the capillary. Increasing the effluent velocity reduces this effect by moving the solutes more rapidly through the thermal gradient into the hot interface. This apparently reduces the band broadening in the connecting capillary and the interface.

Finally, it is important to know the maximum band broadening in the interface and the detector that can be tolerated compared to the band broadening in the column. This is demonstrated by an example. A column with a plate number of 10,000 will give a variance of $36 \, s^2$ for an analyte with a retention time of 10 min. If an external band broadening of $10\%$ (or $\sigma = 0.6 \, s$) is considered to be acceptable, $M_2(s^2)$ has to be lower than $7.6 \, s^2$. As is shown in Table I, this is true for all relatively volatile compounds, and for TPhP and TOCP at a flow-rate of $60 \, \mu l/min$.

## Analytical data

A mixture of TMP, TPP, TBP, TOP, TPhP and TOCP was chromatographed on a C18-bonded phase with methanol-water (80:20) as the eluent. Initially, TEP was also included in the mixture; however, this solute co-eluted with TMP. A typical chromatogram obtained for injected amounts of 50–70 ng of each of the test solutes is shown in Figure 2. The separation was carried out on a 1 mm I.D. column at an eluent flow-rate of $70 \, \mu l/min$, which is about the highest flow-rate that can be handled by the detector without baseline stability problems. Previously, TEP has been shown[8] to have a detection limit of 40 pg of injected compound at an eluent flow-rate of $40 \, \mu l/min$. In this study, comparable detection limits (signal-to-noise ratio, 3:1) were found for most other phosphates, viz. 40–50 pg for TMP, TPP and TBP and 80–100 pg for TPhP and TOCP. The detection limit of TOP was much higher, i.e., about 1 ng. This is at least partly due to the rather high retention of TOP (cf.

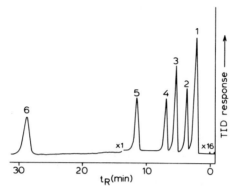

**Figure 2**   LC-TID chromatogram of 6 alkyl and aryl phosphates. Column, glass-lined stainless steel (20 cm × 1 mm I.D.) packed with 5 μm Spherisorb ODS 2. Eluent, methanol-water (80:20) at a flow-rate of 70 μl/min. Solutes: 1, TMP; 2, TPP; 3, TPhP; 4, TBP; 5, TOCP and 6, TOP. Injected amounts, 50–70 ng.

Figure 2; $k' = 14$) and the relatively low weight percentage of phosphorus of this analyte.

The repeatability of the system was determined for injected analyte amounts of 50–70 ng at an LC flow-rate of 35–55 μl/min. At 35 μl/min, the rel. S.D. for TMP, TPP and TBP was between 2.5 and 5.0% ($n = 10$), and the rel. S.D. for TPhP and TOCP were 8.5 and 9.5% ($n = 10$), respectively. At 55 μl/min, the rel. S.D. for TPhP and TOCP improved to 5.5 and 6.0% ($n = 10$), respectively, while the repeatability did not change for the other compounds.

## Determination of triaryl phosphates in sediment

Gas chromatography (GC)-TID[11–13] and GC-mass spectrometry[13] are the methods of choice for the determination of triaryl phosphates in environmental samples (e.g., sediment, fish, water). LC has been applied to the determination of triaryl phosphates, either using rather non-selective UV detection at 260 nm[16] or highly sophisticated graphite furnace atomic absorption detection.[17] Because of this, the use of the thermionic detector is becoming increasingly important, because it can fill the need for a selective and relatively simple LC detector for such compounds.

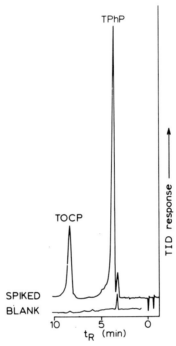

**Figure 3** LC-TID chromatogram of the extract of a sediment spiked with 2 ppm of TPhP and TOCP. Column, glass-lined stainless steel (20 cm × 0.7 mm I.D.) packed with 5 μm Spherisorb ODS 2. Eluent, methanol-water (80:20) at a flow-rate of 35 μl/min. Injected amounts, 20 ng TPhP and 10 ng TOCP. For sample preparation, see Experimental.

A sediment sample from the lake Het Nieuwe Meer (Amsterdam, The Netherlands) was spiked with 2 ppm of each TPhP and TOCP. The sample preparation was limited to a 4 h soxhlet extraction with aqueous methanol, which has been shown to be more efficient than less polar solvents.[11] Figure 3 shows the LC-TID chromatogram of the spiked sediment. Practically no interferences from the sediment matrix are observed in the chromatogram. The high selectivity of the LC-TID system for the aryl phosphates is similar to that previously observed for phosphorus-containing pesticides in tomato, cabbage and onion samples.[7,8] In Reference 7, the selectivity of the LC-TID system was calculated to be $1 \times 10^5$ g of carbon per g of phosphorus.

## CONCLUSIONS

The performance of an interface recently described[8] for on-line LC-TID has been further evaluated. Generally speaking, band broadening in the interface increases with decreasing volatility of the organophosphorus compounds. For the relatively non-volatile phosphates such as TPhP, TOCP and TOP, the volumetric band broadening considerably decreases when the eluent flow-rate is increased. However, the opposite is true for the more volatile test compounds. Optimization will therefore at least partly be determined by the nature of the sample constituents to be analyzed. For the rest, it is important to note that the highest eluent flow-rate that can be handled by the present LC-TID system is 70 $\mu l/min$.

The determination of trialkyl and triaryl phosphates by means of LC-TID is a relevant example, because these compounds lack specific UV absorption and native fluorescence.[10, 16] The simple procedure elaborated for sediment samples containing low-ppm levels of triaryl phosphates illustrates the selectivity and sensitivity of LC-TID in environmental analysis.

The determination of weak organic and inorganic phosphorus-containing acids will be the focus of future work. Preliminary results indicate that acceptable peak profiles can be obtained for, e.g., orthophosphoric acid, provided an eluent having a pH of 2.0–2.8 is used.

### Acknowledgements

Dr. D. Barceló is the recipient of a NATO fellowship (Spain). The authors thank Beun de Ronde (Amsterdam, The Netherlands) for supplying the ISCO equipment. We thank A. Verweij (TNO, Rijswijk, The Netherlands) for the stimulating discussions.

### References

1. U. A. Th. Brinkman, R. B. Geerdink and A. de Kok, *J. Chromatogr.* **291,** 195 (1984).
2. F. A. Maris, R. B. Geerdink and U. A. Th. Brinkman, *J. Chromatogr.* **328,** 93 (1985).
3. A. P. Bruins, *J. Chromatogr.* **323,** 99 (1985).
4. Y. Ito, T. Takeuchi, D. Ishii and M. Goto, *J. Chromatogr.* **346,** 161 (1985).
5. V. L. McGuffin and M. V. Novotny, *Anal. Chem.* **53,** 946 (1981).

6. J. C. Gluckman and M. V. Novotny, *J. Chromatogr.* **333**, 291 (1985).

7. F. A. Maris, R. J. van Delft, R. W. Frei, R. B. Geerdink and U. A. Th. Brinkman, *Anal. Chem.* **58**, 1634 (1986).

8. J. C. Gluckman, D. Barceló, G. J. de Jong, R. W. Frei, F. A. Maris and U. A. Th. Brinkman, *J. Chromatogr.* **367**, 35 (1986).

9. P. H. Howard and P. G. Deo, *Bull. Environm. Contam. Toxicol.* **22**, 337 (1979).

10. D. G. Muir, In: *The Handbook of Environmental Chemistry, Vol. 3, Part C, Anthropogenic Compounds* (O. Hutzinger, ed.) (Springer Verlag, Berlin, 1984), p. 41–66.

11. D. G. Muir, N. P. Grift and J. Solomon, *J. Assoc. Off. Anal. Chem.* **64**, 79 (1981).

12. P. Lombardo and I. J. Egry, *J. Assoc. Off. Anal. Chem.* **62**, 47 (1979).

13. G. L. LeBel, D. T. Williams and F. M. Benoit, *J. Assoc. Off. Anal. Chem.* **64**, 991 (1981).

14. M. N. Krishnamurthy, S. Rajalakshmi and O. P. Kapur, *J. Assoc. Off. Anal. Chem.* **68**, 1074 (1985).

15. J. P. Foley and J. G. Dorsey, *Anal. Chem.* **55**, 730 (1983).

16. E. A. Sugden, R. Greenhalgh and J. R. Pettit, *Environ. Sci. Technol.* **14**, 1498 (1980).

17. P. Tittarelli and A. Mascherpa, *Anal. Chem.* **53**, 1466 (1981).

# Gas Chromatography of Organophosphorus Compounds on Chiral Stationary Phases†

CARLA E. A. M. DEGENHARDT, ALBERT VERWEIJ
and HENK P. BENSCHOP

*Prins Maurits Laboratory TNO, P.O. Box 45, 2280 AA Rijswijk,
The Netherlands*

(*Received October 2, 1986; in final form Novermber 1, 1986*)

The gas chromatographic separation of the stereoisomers of several chiral organo-phosphorus compounds is described. Glass capillary columns coated with the non-chiral phases SE-30 and Carbowax 20M, or with the chiral phases Chirasil-Val and Ni(II)Bis[(1R)-3-(heptafluorobutyryl)camphorate] were used.

KEY WORDS:    Organophosphorus compounds, gas chromatography, stereoisomers, chiral stationary phases.

## INTRODUCTION

Nerve agents are organophosphorus compounds which show strong cholinesterase-inhibiting properties. As a consequence these compounds are extremely toxic. In our institute the pharmacokinetics of nerve agents with the general formula

$$R{-}O \diagdown \qquad \diagup \diagup O$$
$$P_4 *$$
$$Y \diagup \qquad \diagdown X$$

†Presented at the Workshop on Chemistry and Fate of Organophosphorus compounds, Amsterdam, Holland, June 18–20, 1986.
This article was first published in *International Journal of Environmental Analytical Chemistry*, Volume 30, Numbers 1 + 2 (1987).

173

where   $R = $ alkyl   or   $R = C_2H_5$

$\qquad Y = CH_3 \qquad\qquad Y = N(CH_3)_2$

$\qquad X = F \qquad\qquad\quad X = CN$

are studied.

All these organophosphorus compounds have a centre of asymmetry (a chiral centre) at the phosphorus atom, resulting in the existence of enantiomers. As enantiomers may differ in their pharmacokinetic properties, an analytical procedure had to be developed which would allow separate determination of the isomers. In this paper the application of gas chromatography with chiral and non-chiral capillary columns for the separation of enantiomeric and diastereoisomeric organophosphorus compounds will be shown.

## EXPERIMENTAL PART

Soman or 1,2,2-trimethylpropyl methylphosphonofluoridate (Figure 1) has been the principal test compound. Having chiral centres both at the P-atom and at the alpha-C-atom, four stereoisomers of Soman exist, designated as $C(-)P(+)$, $C(+)P(-)$, $C(+)P(+)$ and $C(-)P(-)$ in which C stands for the alpha-C-atom in the 1,2,2-trimethylpropylgroup and P for the phosphorus atom.

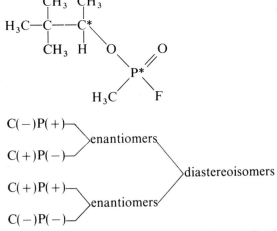

**Figure 1**   Structure of 1,2,2-trimethylpropyl methylphosphonofluoridate (Soman).

The four isomers show widely differing toxicological properties as published by Benschop et al.[1-4]

The diastereoisomeric pairs, having different physico-chemical properties, can be separated on non-chiral stationary phases. However, the enantiomers, having identical physico-chemical properties can only be separated on chiral stationary phases.

## GLC of diastereoisomers on non-chiral stationary phases

In 1971 Verweij et al.,[5] published the gas chromatographic separation of a number of diastereoisomeric alkyl methylphosphono-fluoridates and related compounds on polar and apolar packed columns. Among other structural aspects it was demonstrated that the relative retention ($r$) i.e. the ratio of the adjusted retention times of the diastereoisomers increased both on lengthening and on branching the alkyl chain as well as on applying a polar column instead of an apolar column. With Soman the experiment was repeated on capillary columns, giving comparable relative retention values; the chromatograms are presented in Figure 2.

The ($+$) and ($-$) designations for C($\pm$)P($\pm$) Soman concerning the configuration were derived from reference experiments with synthesized or isolated optically pure compounds.[4] The peak area

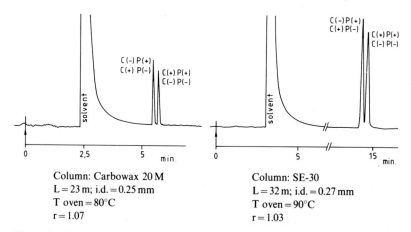

Column: Carbowax 20 M
L = 23 m; i.d. = 0.25 mm
T oven = 80°C
r = 1.07

Column: SE-30
L = 32 m; i.d. = 0.27 mm
T oven = 90°C
r = 1.03

**Figure 2** GLC of the diastereoisomers of Soman on Carbowax 20 M and SE-30 capillary columns.

ratio of the diastereoisomers in synthesized $C(\pm)P(\pm)$ Soman proved to be constantly amounting to 55:45.

As can be seen from Figures 5 and 6 each diastereoisomeric pair contains equal amounts of the respective enantiomers.

## GLC of stereoisomers on the Chirasil-Val chiral stationary phase

In 1978 Frank *et al.*,[6] introduced the chiral gas chromatographic phase Chirasil-Val consisting of L-valine-tert.-butylamide bonded to a polysiloxane skeleton (Figure 3).

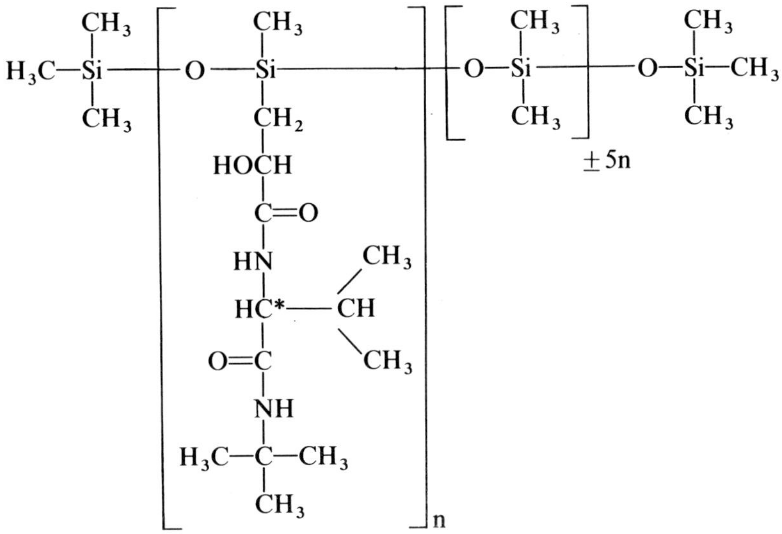

**Figure 3**   Structure of Chirasil-Val.

In our laboratory the synthesis of Chirasil-Val led to two different qualities of Chirasil-Val, which were coded type I and type II. They show different GC separation capabilities for Soman as was found in experiments on the coated capillary columns. Until recently the cause for this difference in behaviour was unknown. However, now we know which, obviously, critical step in the synthesis of Chirasil-Val is responsible for the encountered differences. The separation of

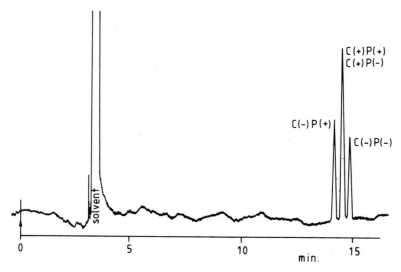

**Figure 4**   GLC of the stereoisomers of Soman on Chirasil-Val type I.
L = 48 m, i.d. = 0.44 mm, T oven = 75°C.

Soman isomers on Chirasil-Val type I is presented in Figure 4. Only three peaks were obtained which could be designated as C($-$)P($+$), C($+$)P($-$)+C($+$)P($+$), C($-$)P($-$) respectively. This chromatogram shows the ability of this chiral phase to separate the enantiomers.

As diastereoisomers can be separated on non-chiral columns, the Chirasil-Val type I column was extended with a Carbowax 20 M column, thus giving complete separation of the four Soman stereoisomers as can be seen in Figure 5. This chromatogram demonstrates again the typical peak area ratio amounting to 55:45 of the diestereoisomers. The peak area of the enantiomers are identical.

The separation of Soman isomers on Chirasil-Val type II is presented in Figure 6. This type of Chirasil-Val is able to separate Soman into four peaks without the assistance of an additional non-chiral column.

As can be derived from Figures 5 and 6 the elution order of the two inner peaks is inversed. This inversion is confirmed by the characteristic peak ratios of the diastereoisomers as well as by injection of the optically pure Soman components.

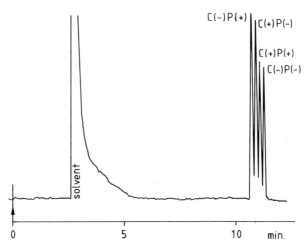

**Figure 5**   GLC of stereoisomers of Soman on Chirasil-Val type I—Carbowax 20 M
combination.
Chirasil-Val: L = 48 m, i.d. = 0.44 mm
Carbowax 20 M: L = 14 m, i.d. = 0.48 mm
T oven = 80 °C.

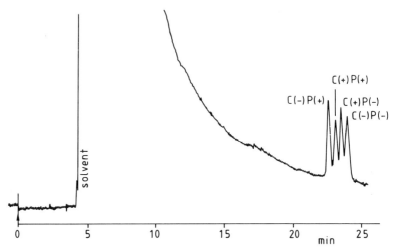

**Figure 6**   GLC of the stereoisomers of Soman on Chirasil-Val type II.
L = 50 m, i.d. = 0.5 mm, T oven = 80 °C.

Besides Soman, the enantiomers of isopropyl methylphosphono-fluoridate (Sarin) and cyclohexyl methylphosphonofluoridate (cyclo-hexylsarin) were resolved as presented in Figure 7.

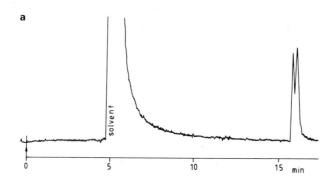

isopropyl methylphosphonofluoridate
(Sarin)

cyclohexyl
methylphosphonofluoridate

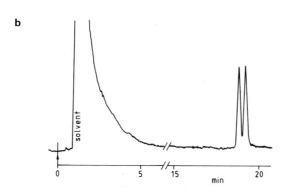

**Figure 7** GLC of the enantiomers of Sarin (Figure 7a, T oven = 60°C) and of cyclohexylsarin (Figure 7b, T oven = 90°C) on Chirasil-Val type II. L = 28 mm, i.d. = 0.25 mm

The preparation and deactivation of the GLC columns have been described by Benschop *et al.*[7]

Proper deactivation of the Chirasil-Val columns proved to be important as can be seen from Figure 10.

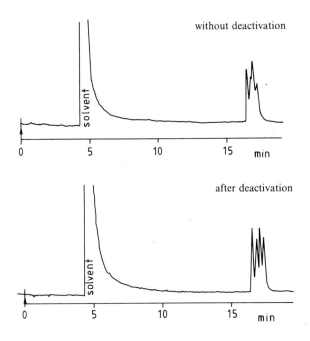

**Figure 10**   Effect of deactivation on the separation of the stereoisomers of Soman on Chirasil-Val type II.
L = 50 m, i.d. = 0.5 mm, T oven = 90°C.

Additionally some commercially available Chirasil-Val-coated columns were tried, but none of these columns was able to separate Soman in more than three peaks. They proved to correspond with the Chirasil-Val type I-coated columns prepared in our laboratory.

## GLC of deuterated internal standards

The quantitative GLC analysis of trace amounts of Soman stereoisomers in biological material, in which these isomers are highly labile, necessitates the use of internal standards with properties similar to those of the Soman stereoisomers. So, deuterated Soman stereoisomers were synthesized[3] in house, coded $[^2H_3]$-Soman (deuterated P-$CH_3$ group) or $[^2H_{13}]$-Soman (deuterated 1,2,2-trimethylpropyl group).

As the elution order of the Soman stereoisomers depends on the quality of the synthesized Chirasil-Val, the internal standard had to be adapted to the type of Chirasil-Val used.[7]

This means that, if the GC analyses were to be carried out on

a) a Chirasil-Val type I column combined with a Carbowaz 20 M column, then the internal standard used was $C(-)P(+)-[^2H_3]$ Soman,

b) a Chirasil-Val type II column, then the internal standards used were $C(\pm)P(+)-[^2H_{13}]$ Soman.

Figures 8 and 9 show the respective separation of the Soman stereoisomers and the internal standards on the two different Chirasil-Val types.

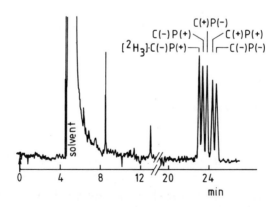

**Figure 8** GLC of $C(-)P(+)[^2H_3]$ Soman and the stereoisomers of Soman on Chirasil-Val type I—Carbowax 20 M combination.
Chirasil-Val: $L = 48$ m, i.d. $= 0.44$ mm
Carbowax 20 M: $L = 14$ m, i.d. $= 0.48$ mm
T oven $= 80°$C.

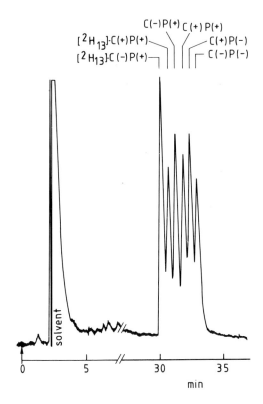

**Figure 9**   GLC of the stereoisomers of $C(\pm)P(+)[^2H_{13}]$ Soman and of Soman on Chirasil-Val type II.
L = 50 m, i.d. = 0.5 mm T oven = 80°C.

## GLC of stereoisomers on chiral complexation columns

Since 1977 Schurig *et al.*, have published[8-10] the separation of several classes of nonphosphorus chiral compounds by complexation gas chromatography on metal complexes with Ni(II)Bis[(1R)-3-(heptafluorobutyryl)camphorate] (Figure 11) as an efficient chiral stationary phase.

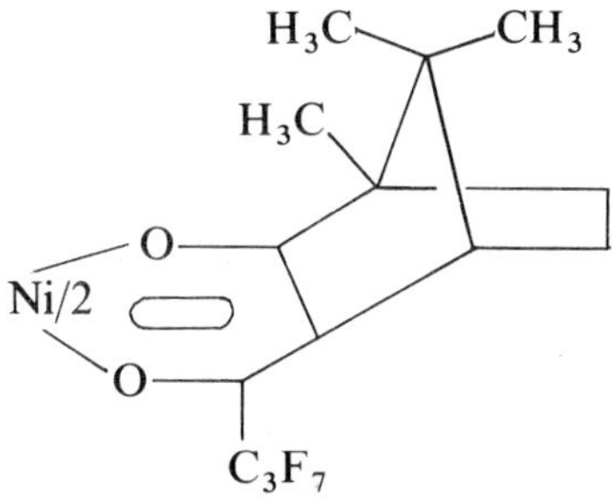

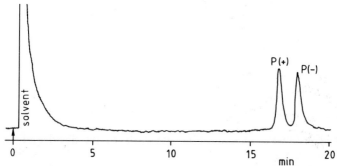

**Figure 11** Structure of the chiral metal chelate Ni(II)Bis[(IR)-3-(heptafluorobutyryl) camphorate].

A capillary column coated with this phase in OV-101 separates the enantiomers of O-ethyl N,N-dimethylphosphoramidocyanidate (Tabun) as shown in Figure 12.[11] On Chirasil-Val columns (type I and type II) the enantiomers coincide.

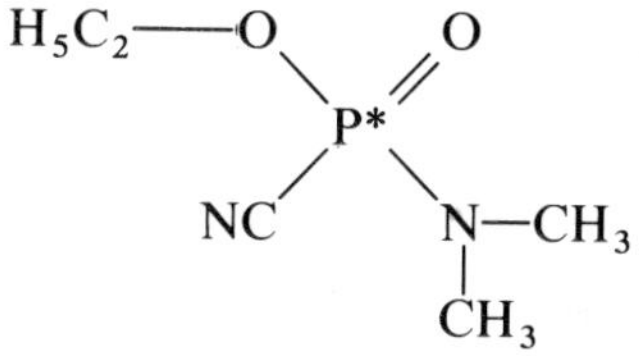

O-ethyl N,N-dimethylphosphoramidocyanidate
(Tabun)

**Figure 12** Separation of Tabun enantiomers on an OV-101/Ni(HFB-1R-Cam)$_2$ complexation column.
L = 14 m, i.d. = 0.44 mm, T oven = 120°C

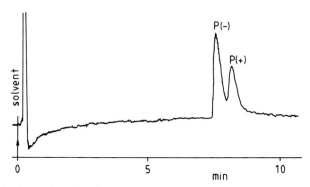

**Figure 13**  Separation of Sarin enantiomers on an OV-101/Ni(HFB-1R-Cam)$_2$ complexation column.
L = 9 m, i.d. = 0.44 mm, T oven = 100°C.
An excess of the ( − ) Sarin had been added to a racemic mixture.

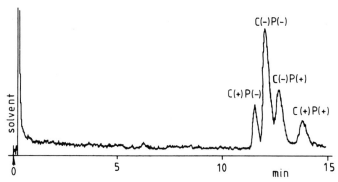

**Figure 14**  Separation of Soman stereoisomers on an OV-101/Ni(HFB-1R-Cam)$_2$ complexation column.
L = 9 m, i.d. = 0.44 mm, T oven = 120°C.
An excess of C( − )P( − ) Soman had been added to C( ± )P( ± ) Soman.

Figures 13 and 14 show the separation of the Sarin and Soman stereoisomers respectively.

The elution order of the Soman stereoisomers on the complexation phase differs from that on the Chirasil-Val phase. On a Chirasil-Val column (type I and type II) the first peak belongs to C( − )P( + )

**Table I**  Separation capability of some gas chromatographic stationary phases for Soman

| Stationary phase | Separation capability, number of peaks | Elution order | | | |
|---|---|---|---|---|---|
| SE-30; Carbowax 20 M | 2 | C(−)P(+) | C(+)P(+) | | |
| | | C(+)P(−) | C(−)P(−) | | |
| Chirasil-Val type I | 3 | C(−)P(+) | C(+)P(+) | C(−)P(−) | C(+)P(−) |
| Chirasil-Val type I +Carbowax 20 M | 4 | C(−)P(+) | C(+)P(−) | C(+)P(+) | C(−)P(−) |
| Chirasil-Val type II | 4 | C(−)P(+) | C(+)P(+) | C(+)P(−) | C(−)P(−) |
| OV-101/Ni (HFB-1R-Cam)$_2$ | 4 | C(+)P(−) | C(−)P(−) | C(−)P(+) | C(+)P(+) |

Soman, whereas on the complexation column the first peak belongs to its enantiomer $C(+)P(-)$ Soman.

The synthesis of the metal chelate and the preparation of the glass capillary columns were carried out in our laboratory according to Schurig *et al.*[9,10]

## SUMMARY

Table I summarizes gas chromatographic separation of some chiral organophosphorus compounds on two different chiral stationary phases and on polar and apolar non-chiral stationary phases. The results presented are focussed on the elution order of the peaks of the four stereoisomers of Soman.

### Acknowledgement

The authors gratefully acknowledge George R. van den Berg and Matthijs F. Otto for the Synthesis of Chirasil-Val, and Roland P. E. van Damme and Ger W. H. Moes for the synthesis of the nickel chelate.

### References

1. H. P. Benschop, F. Berends and L. P. A. de Jong, *Fundam. Appl. Toxicol.* **1**, 177 (1982).
2. H. P. Benschop, C. A. G. Konings and L. P. A. de Jong, *J. Amer. Chem. Soc.* **103**, 4260 (1981).
3. H. P. Benschop, J. van Genderen and L. P. A. de Jong, *Toxicol. Appl. Pharmacol.* **72**, 61 (1984).
4. H. P. Benschop, C. A. G. Konings, J. van Genderen and L. P. A. de Jong, *Fundam. Appl. Toxicol.* **4**, 84 (1984).
5. A. Verweij, E. Burghardt and A. W. Koonings, *J. Chromatogr.* **54**, 151 (1971).
6. H. Frank, G. J. Nicholson and E. Bayer, *Angew. Chem.* **90**, 396 (1978).
7. H. P. Benschop, E. C. Bijleveld, M. F. Otto, C. E. A. M. Degenhardt, H. P. M. van Helden and L. P. A. de Jong, *Anal. Biochem.* **151**, 242 (1985).
8. V. Schurig, *Chromatographia* **13**, 263 (1980).
9. V. Schurig, W. Bürkle, *J. Amer. Chem. Soc.* **104**, 7573 (1982).
10. V. Schurig and R. Weber, *J. Chromatogr.* **289**, 321 (1984).
11. C. E. A. M. Degenhardt, G. R. van den Berg, L. P. A. de Jong, H. P. Benschop, J. van Genderen and D. van der Meent, to be published.

# Assay of the Chiral Organophosphate, Soman, in Biological Samples[†]

## L. P. A. DE JONG, E. C. BIJLEVELD, C. VAN DIJK and H. P. BENSCHOP

*Prins Maurits Laboratory TNO, P.O. Box 45, 2280 AA Rijswijk, The Netherlands*

(*Received June 15, 1986; in final form September 22, 1986*)

The anticholinesterase, soman, $(CH_3)_3CC(H)CH_3O(CH_3)P(O)F$, consists of four stereoisomers assigned as $C(\pm)P(\pm)$-soman in which C stands for chirality in the pinacolyl moiety and P for chirality at phosphorus. The four stereoisomers are separated by gas chromatography on an optically active Chirasil–Val column, synthesized and coated in house, or on a Chirasil–Val column identical with the commercially available column when combined with a Carbowax 20M column. This method in combination with an assay based on acetylcholinesterase inhibition shows that the two isomers which do not have anticholinesterase activity, i.e. $C(\pm)P(+)$-soman, are rapidly degraded in rat blood due to hydrolysis by phosphoryl-phosphatases. Epimeric soman isomers, e.g. $C(\pm)P(-)$-soman, can be separately assayed on a Carbowax or a CPSil 8 column, using $^2H$-labeled soman isomers as internal standards. $^2H$-labeled soman stereoisomers serve as internal standards in GC-assay of all four stereoisomers on Chirasil–Val.

For work-up of the four stereoisomers from rat blood the sample is first stabilized by (i) acidification to pH 4.2 at 0°C to suppress hydrolysis by phosphoryl-phosphatases, (ii) addition of aluminum ions for complexation of fluoride ions to prevent regeneration of $C(\pm)P(-)$-soman by free fluoride ions from soman-inhibited carboxylesterase, and (iii) addition of $(CH_3)_3CCH_2O(CH_3)P(O)F$ to occupy covalent binding sites for $C(\pm)P(-)$-soman, before extraction with a Sep-Pak $C_{18}$ cartridge and elution with ethyl acetate. Using a splitless or on-column injection technique and

---

†Presented at the Workshop of Chemistry and Fate of Organophosphorus Compounds, Amsterdam, June 18–20, 1986.

This article was first published in *International Journal of Environmental Analytical Chemistry*, Volume 29, Number 3 (1987).

alkali flame ionization detection, the minimum detectable concentration is 30 pg/3-ml blood sample.

KEY WORDS:    Soman, stereoisomers, gas chromatography, metabolism, rat blood.

## INTRODUCTION

The toxicity of the nerve agent, soman, $(CH_3)_3CC(H)CH_3O(CH_3)P(O)F$ (see Figure 1), is mainly based on its high inhibitory potency for acetylcholinesterase (AChE). Treatment of soman intoxication with the standard antidotes against anticholinesterases, atropine and an oxime, is hampered by rapid conversion of soman-inhibited AChE into a form which can no longer be reactivated by oxime (aging).[1,2] In addition, evidence has accumulated for a much longer *in vivo* persistence of the agent than previously assumed, presenting further difficulties in medical treatment of soman intoxications.[3] Toxico-kinetic studies for assessment of the role of intact soman in the body require methods for analysis of trace amounts of this highly toxic agent. Such studies are complicated by the presence of two chiral centers in the soman molecule leading to four stereoisomers desig-nated as $C(+)P(+)$, $C(+)P(-)$, $C(-)P(+)$ and $C(-)P(-)$ in which C and P stand for the two chiral centers (Figure 1). These isomers may differ in inhibitory potency towards AChE[4] and other esterases,[5] acute toxicity,[6] rate of detoxification,[7,8] rate of passage through

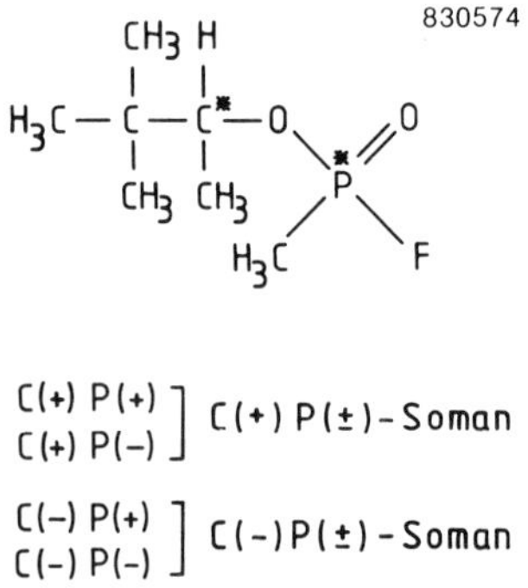

**Figure 1**   The four stereoisomers of $C(\pm)P(\pm)$-soman. The chiral centers are denoted by an asterisk.

membranes, reversible binding, etc. Analysis methods should, therefore, distinguish between the four stereoisomers.

In this paper studies will be described in which GC-analysis methods were developed for this purpose. In addition, the paper deals with a study on the development of a work-up procedure of the four stereoisomers from rat blood samples in toxicokinetic studies.


## MATERIALS AND METHODS

### Materials

Electric eel acetylcholinesterase (AChE, EC 3.1.1.7) and bovine pancreas chymotrypsin (EC 3.4.21.1) were procured from Sigma Chemical Co., St. Louis, U.S.A., as preparations type V-S or VI-S and preparation type II, respectively.

The organophosphates† were prepared in this laboratory according to standard procedures. (+)- and (−)-pinacolyl alcohol, starting compounds for synthesis of $C(+)P(\pm)$-soman and $C(-)P(\pm)$- and $D_3$-$C(-)P(\pm)$-soman, were obtained by resolution of the alcohol as described by Benschop et al.[9] The deuterated starting compounds for the synthesis of $D_{13}$- and $D_3$-soman, $D_{13}$-pinacolyl alcohol and $D_3$-methylphosphonic dichloride and difluoride, respectively, were prepared from $D_6$-acetone and $D_3$-methyl iodide, respectively.[10] The single stereoisomers of soman as well as $D_3$-$C(-)P(+)$-soman and $D_{13}$-$C(\pm)P(+)$-soman were isolated after treatment of the appropriate preparation with chymotrypsin ($P(+)$-isomers) or rabbit serum ($P(-)$-isomers) according to Benschop et al.[9,10] The synthesis of Chirasil–Val was performed according to Bayer and Frank[11-13] except for the final condensation step of Chirasil with L-valine tert.-butylamide which was performed with N-ethoxycarbonyl-2-ethoxy-1,2-dihydroquinoline instead of N,N′-dicyclohexylcarbodiimide.[10] All other chemicals were obtained commercially.

---

†Warning: In view of their extreme toxicities, the organophosphates should be handled only in specialized laboratories, where trained medical personel is continuously present.

## Coating of capillary columns

Capillaries drawn from Duran 50 glass were roughened by deposition of NaCl and coated dynamically with a 25% solution of Chirasil–Val in *n*-pentane or a solution of Carbowax 20M in methylene chloride by using the mercury plug method.[10,14,15]

## Gas chromatography

A gas chromatograph (Packard model 428, Carlo Erba 5160, Pye-104 or Perkin-Elmer Sigma 3) equipped with an alkali flame ionization detector or a FID-detector was used. The following glass capillary columns were applied: type I Chirasil–Val[10] (coated in house 1 = 48 m, i.d. = 0.5 mm)/Carbowax 20M (coated in house, 1 = 14 m, i.d. = 0.5 mm), type II Chirasil–Val[10] (coated in house, 1 = 50 m, i.d. = 0.5 mm) and Chirasil–Val (Applied Science, State College, PA16801, U.S.A., 1 = 25 m, i.d. = 0.3 mm)/Carbowax 20M (coated in house, 1 = 30 m, i.d. = 0.3 mm). The fused silica columns with chemically bonded phase were Carbowax 57 CB (Chrompack, The Netherlands, 1 = 50 m, i.d. = 0.32 mm) and CPSil 8 CB (Chrompack, The Netherlands, 1 = 51 m, i.d. = 0.32 mm, film thickness 1.3 $\mu$m). The Chirasil–Val and Chirasil–Val/Carbowax columns were operated at 80°C,[10,16] whereas for chromatographic runs on Carbowax 57 CB or CPSil 8 CB the columns were heated from 87–104°C at a rate of 1°C/min or from 80–87°C at 50°C/min and from 87–140°C at 6.2°C/min, respectively. Carrier gas was helium (1–2 ml/min).

## Determination of residual soman concentration after degradation in rat blood

Supernatants of blood or plasma samples mixed with a 3-fold or 1.5-fold volume, respectively, of formate buffer (25 mM, pH 3.75) and centrifuged (20,000 g, 10 min), were assayed both enzymatically and gas chromatographically. In the enzymatic assay the supernatants were incubated with AChE for a fixed time (5 min). From the percentage of residual enzyme activity the soman concentration was evaluated with the aid of a calibration curve made with $C(\pm)P(\pm)$-soman.[17] In the GC-assay the supernatants were passed through a XAD-2 column which was subsequently eluted with ethyl acetate. After concentration at reduced pressure (7–8 kPa) to approximately

0.1 ml, the eluate was analyzed on a glass capillary SE-30 column $(1 = 35\,m,\ i.d. = 0.7\,mm)$, operated at 130°C with helium $(5\,ml/min)$ as carrier gas. The soman analogue $(CH_3)_2CHC(H)CH_3O(CH_3)P(O)F$ was used as an internal standard.

## Enzyme assays

AChE was assayed titrimetrically (pH 7.5, 25°C) with acetylcholine perchlorate as a substrate. Carboxylesterase activities were determined both titrimetrically (pH 7.5, 25°C) and spectrophotometrically (pH 7.0, 25°C) with methyl butyrate (0.06 M) and $o$-nitrophenyl acetate (0.67 mM), respectively, as substrates.[17]

## Work-up procedure for GC determination of the soman stereoisomers

Blood samples were mixed with a threefold volume of a 0.2 M acetate buffer (pH 3.5) containing 0.82 g saponine/1, 1.65 mM aluminum sulphate and a 100-fold excess of $(CH_3)_3CCH_2O(CH_3)P(O)F$ relative to the expected amount of $C(\pm)P(\pm)$-soman. After addition of internal standard the mixture is pressed through a Sep-Pak $C_{18}$ cartridge. The analytes were eluted with ethyl acetate, after which the eluate was concentrated at reduced pressure (7.3 kPa).

## Animal experiments

Male Wistar (WAG/Rij) rats, weighing 180–200 g and bred in the Medical Biological Laboratory TNO under SPF conditions, were used. For *in vitro* experiments blood was taken by heart puncture under nembutal anesthesia and collected in a syringe containing heparin (1000 IE).

Blood used for the development of the work-up procedure was obtained from rats anesthesized with sodium barbital (200 mg/kg, ip) and sodium hexobarbital (175 mg/kg, ip) and treated with atropine sulphate (50 mg/kg, ip) 5 min before administration of 1 LD50 (82 μg/kg) or 6 LD50 of $C(\pm)P(\pm)$-soman intravenously in the dorsal penis vein. The rats were kept alive with artificial respiration. After 1 h blood was taken via a carotid cannula.

## RESULTS AND DISCUSSION

### Qualitative analysis of soman stereoisomers and applications

*GC separation of the four stereoisomers of soman*   A GC separation of the enantiomers of C($\pm$)P($\pm$)-soman was first achieved on a commercial column coated with the chiral phase Chirasil–Val, a copolymeric organosiloxane bound to L-valine tert.-butylamide. At optimal conditions, C($\pm$)P($\pm$)-soman was resolved in three instead of four peaks. The elution patterns of C($+$)P($\pm$)-soman and C($-$)P($\pm$)-soman showed that the enantiomeric pairs of soman are separated but that the retention times of the C($+$)P($\pm$)-epimers are identical (Figure 2). Resolution of all four stereoisomers was

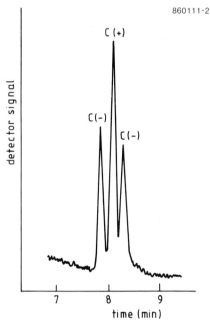

**Figure 2**   GC separation of the soman enantiomers on a commercial Chirasil–Val column at 80°C, precolumn pressure 103 kPa. The injection and FID-detector block of a Pye-104 gas chromatograph are held at 300°C. A 1-$\mu$l sample of 4 mM C($\pm$)P($\pm$)-soman solution in isopropanol is injected (split ratio 1:10). Reprinted with permission from *J. Am. Chem. Soc.* **103**, 4260 (1981). Copyright (1981) American Chemical Society.

obtained with a system consisting of the Chirasil–Val column coupled to a Carbowax 20M column,[16] which resolves[18] the epimers of $C(\pm)P(\pm)$-soman (Figure 3).

These results prompted us to synthesize Chirasil–Val and to coat glass capillary columns with the chiral phase obtained. The quality of the synthesized Chirasil–Val was found to vary between two resolution patterns. Columns coated with type I Chirasil–Val showed properties similar to the commercially available column and when combined with a Carbowax 20M column, separated the four stereo-isomers as shown in Figure 3. On a column coated with type II Chirasil–Val the four stereoisomers are fully separated. Identification of the peaks of the chromatograms (see: *GC identification of anti-cholinesterase stereoisomers of soman*) indicates that the enantiomers are better separated on the type II Chirasil–Val column. The elution order of the enantiomeric pairs are peaks $1+2$ and peaks $2+3$ on type I Chirasil–Val, peaks $1+2$ and peaks $3+4$ on type I Chirasil–

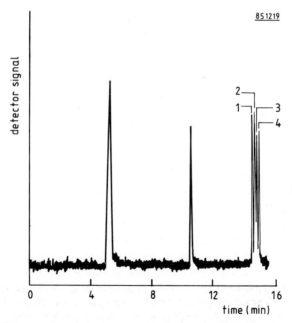

**Figure 3**   GC separation of the four stereoisomers of $C(\pm)P(\pm)$-soman on Chirasil–Val/Carbowax 20M. The Carbowax leg was connected with the injection port of the gas chromatograph. See Figure 2 for further data.

Val/Carbowax, whereas on type II Chirasil–Val the enantiomeric pairs are eluted as peaks $1+3$ and peaks $2+4$. It should be remarked that other investigations[19] also found variation in the chiral resolution quality of Chirasil–Val per synthesized batch.

*GC identification of anticholinesterase stereoisomers of soman*    From kinetic analysis of the inhibition of bovine erythrocyte AChE with equimolar concentration of C($+$)P($\pm$)- or C($-$)P($\pm$)-soman Keijer and Wolring[4] showed that only one of the two stereoisomers of each preparation is a potent inhibitor of AChE. Accordingly, two peaks had almost disappeared in the elution pattern of the residual soman extracted from an incubation mixture of electric eel AChE (1.4 $\mu$M of active sites) with C($\pm$)P($\pm$)-soman. This stereospecific enzyme inhibition was applied to fully identify the peaks in the chromatograms. For that purpose experiments were carried out with chymotrypsin as an enzyme which can be used in a sufficiently high concentration (ca 1.8 mM of active sites) to determine the optical rotation of the residual soman after extraction. Upon incubation with a twofold molar excess of C($+$)P($\pm$)- as well as of C($-$)P($\pm$)-soman the residual soman had a significantly more positive optical rotation than before incubation.[16] GC analysis showed that chymotrypsin and AChE are preferentially inhibited by the same stereoisomers, now identified as the P($-$)-epimers. So, the stereoisomers of soman corresponding to peak 1–4 of Figure 3 (Chirasil–Val/Carbowax) are the C($-$)P($+$)-, C($+$)P($-$)-, C($+$)P($+$)- and C($-$)P($-$)-isomers, respectively.

*Stereospecific degradation of soman in rat blood*    GC analysis of the four stereoisomers was first applied to study the fate of C($\pm$)P($\pm$)-soman in rat blood and plasma *in vitro*. The concentration of each stereoisomer is the resultant of chemical hydrolysis, enzymatic hydrolysis by phosphorylphosphatase, irreversible binding to cholinesterases and other esterases and binding to other sites. After incubation for a relatively short time (10 min) only a small amount of the organophosphate is left as determined both enzymatically from its inhibitory effect on AChE and gas chromatographically on an achiral SE-30 column (Table I).

Soman concentrations were determined after adjusting the pH of the samples to 5.5 (0°C) to preclude further enzymatic degradation

**Table I** Residual soman concentrations[a] in rat blood (A) and plasma (B) incubated with 1.4 and 2.3 μg C(±)P(±)-soman/ml for 10 min at 37°C, subsequently mixed with formate buffer (final pH 5.5) and centrifuged

| Incubate | Treatment with NaF | Enzyme assay C(±)P(−)-soman (μg/ml) | GC assay total soman (μg/ml) | Chiral GC, peak ratios | | | |
|---|---|---|---|---|---|---|---|
| | | | | C(+)P(+) | C(+)P(−) | C(−)P(+) | C(−)P(−) |
| A | none | 0.095±0.009 | 0.082±0.016 | −[b] | 1[c] | − | 1.3 |
| | 0.5 min | 0.20±0.01 | n.d.[d] | | n.d. | | |
| | 30 min | 0.21±0.02 | 0.42±0.03 | 0.8 | 1 | 0.8 | 0.9 |
| B | none | 0.19±0.02 | 0.18±0.04 | − | 1 | − | 1.3 |
| | 0.5 min | 0.38±0.03 | n.d. | | n.d. | | |
| | 30 min | 0.31±0.01 | 0.67±0.13 | 0.9 | 1 | 0.9 | 0.9 |

[a]Mean values and their standard deviations from three separate runs, calculated as concentrations in blood or plasma.
[b]Not detected.
[c]Arbitrarily set to 1.
[d]Not determined.

by phosphorylphosphatase.[7] In the enzymatic assay only the $C(\pm)$ $P(-)$-epimers are determined because of the high reaction rate of these isomers with AChE in comparison with the $C(\pm)P(+)$-epimers. The correspondence between the data obtained in the enzymatic and the GC assays indicates that only $C(\pm)P(-)$-epimers are present in blood and plasma even after a short incubation.[17] This finding based on quantitative analysis was confirmed by qualitative GC assay on a Chirasil–Val/Carbowax column (Table I).

As a control experiment racemization was induced by incubation (30 min) with fluoride ions (2.5 mM). Indeed, the $C(\pm)P(-)$-soman concentration determined enzymatically becomes about half the total soman concentration as determined by achiral GC assay and four peaks with approximately the same height are observed in chiral GC analysis. Surprisingly, the treatment with fluoride ions at pH 5.5 highly increases the soman concentration. This increase is a fast process as appears from the considerable effect brought about within 0.5 min of treatment. Soman formed in this short period consists of $C(\pm)P(-)$-epimers as follows from chiral GC analysis after rapid extraction with ethyl acetate. Evidence is presented for parallel reactivation induced by fluoride ions of carboxylesterase which had been inhibited by soman. The percentage of reactivation after 0.5 min of fluoride treatment (pH 4.8–6.1) is found proportional to the $C(\pm)P(-)$-soman increase brought about under the same conditions. These results strongly indicate that added fluoride ions regenerate $C(\pm)P(-)$-soman by reversal of the inhibition reaction:

where E—OH represents the carboxylesterase.

From the relationship between percentage of reactivation and increase of soman concentration induced by fluoride ions at various conditions (pH, fluoride concentration) the carboxylesterase concentration in rat plasma was extrapolated as $2.6\,\mu M$ corresponding closely to values reported in literature.[20,21] In this way an aspecific binding site contributing substantially to the fate of soman in rat blood was quantified.

The $P(+)$-epimers, with a low anticholinesterase activity, are much more rapidly degraded in rat blood than the $P(-)$-epimers which have a high anticholinesterase activity. Hence, it should be kept in mind that initial rates of overall degradation of soman as usually determined to characterize an enzyme reaction, are probably not relevant for the rate of detoxification of $C(\pm)P(-)$-soman in rat blood.[22,23]

*Isolation of the four stereoisomers of soman*   The availability of a method for the separate analysis of the four stereoisomers of soman meets one of the requirements for their isolation as single isomers. The observed stereospecific enzymatic reactions offer the possibility for selective removal of stereoisomers. So, starting from $C(+)P(\pm)$- and $C(-)P(\pm)$-soman the two $P(+)$- and the two $P(-)$-isomers were isolated after incubation with chymotrypsin and in rabbit serum, respectively. At optimal conditions, the stereoisomers were obtained with more than 99% optical purity[9] on a 1 to 15 mg scale which is sufficient for toxicological investigations.

## Quantitative analysis of soman stereoisomers and applications

*Quantitative GC assay of the soman stereoisomers*   Quantitative GC assay of trace amounts of soman stereoisomers in biological material, in which these isomers may be highly labile, necessitates the use of internal standards whose properties are very similar to those of soman. We investigated the feasibility of two deuterated isomers of soman as isotopically labeled internal standards: $D_3$-$C(\pm)P(\pm)$-soman and $D_{13}$-$C(\pm)P(\pm)$-soman, in which the three hydrogen atoms of the P—$CH_3$ group and all hydrogen atoms of the pinacolyl moiety, respectively, are replaced by deuterium (Figure 4).

As the $P(+)$-epimers rapidly disappear in rat blood, separate

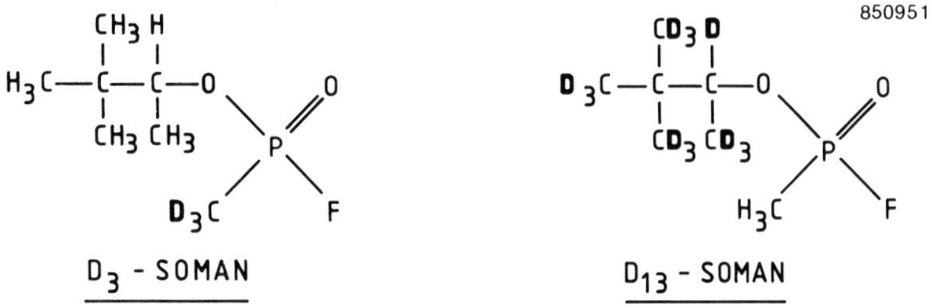

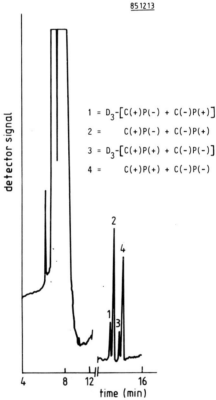

850951

$D_3$ - SOMAN

$D_{13}$ - SOMAN

**Figure 4** Chemical structures of partly deuterated soman analogues, for use as internal standards.

851213

1 = $D_3$-[C(+)P(-) + C(-)P(+)]

2 =     C(+)P(-) + C(-)P(+)

3 = $D_3$-[C(+)P(+) + C(-)P(-)]

4 =     C(+)P(+) + C(-)P(-)

**Figure 5** GC separation of the enantiomeric pairs of C($\pm$)P($\pm$)-soman and of $D_3$-C($\pm$)P($\pm$)-soman on a Carbowax 57 CB fused silica column. Cold on-column injection of a 3-$\mu$l sample in ethyl acetate with 10 s secondary cooling time. Alkali flame ionization detection. See Materials and Methods for further details.

assay of the residual P(−)-epimers can *a priori* be carried out on an achiral column, as already mentioned. We developed assays for these isomers on the basis of their resolution on a Carbowax column and on a SE-54 (CPSil 8) column. As shown in Figure 5 the enantiomeric pairs of soman and of $D_3$-soman are well separated on a Carbowax column, whereas on a SE-54 column the separate assay of the P(−)-epimers can well be performed by using $D_{13}$-soman as an internal standard (Figure 6).

The deuterated soman-isomers as such cannot be used as internal standards for the quantitative assay of all four stereoisomers of soman on type I or II Chirasil–Val columns, since only one or two of the deuterated stereoisomers are eluted separately from the soman stereoisomers. As $D_3$-C(−)P(+)-soman and the $D_{13}$-C(±)P(+)-

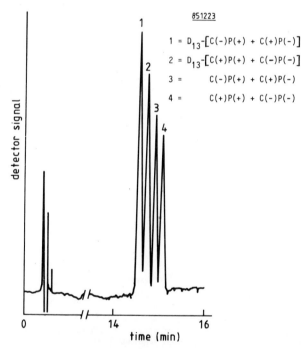

**Figure 6**  GC separation of the enantiomeric pairs of C(±)P(±)-soman and of $D_{13}$-C(±)P(±)-soman on a SE-54 (CPSil 8 CB) fused silica column. On-column injection of a 3-$\mu$l sample in ethyl acetate with 10 s secondary cooling time. Alkali flame ionization detection. See Materials and Methods for further details.

epimers are fully resolved from the soman stereoisomers on a type I and a type II Chirasil–Val column, respectively, these isomers were isolated by analogy with the isolation of the non-deuterated soman stereoisomers, starting from $D_3$-C($-$)P($\pm$)-soman, prepared from ($-$)-pinacolyl alcohol, and from $D_{13}$-C($\pm$)P($\pm$)-soman, respectively. These preparations serve as convenient isotopically and stereo-chemically labeled internal standards for the quantitative assay of the four stereoisomers of soman (Figures 7 and 8).

*Procedure for work-up of the four stereoisomers of soman from rat blood* A quantitative determination of the stereoisomers of soman requires a work-up procedure during which the labile soman isomers are stabilized. On the basis of the work of Christen[7] we tried to block the activity of phosphorylphosphatase being especially active in hydrolysis of the P($+$)-epimers, by acidification of the blood sample with a threefold volume of 0.2 M acetate buffer (pH 3.5). To check the stability of soman we used blood from anesthesized, atropinized and artificially respirated rats which had received 1 or 6 LD50 C($\pm$)P($\pm$)-soman 1 h earlier. In this way we obtained blood samples having their irreversible sites occupied as well as very low levels of residual C($\pm$)P($-$)-soman.

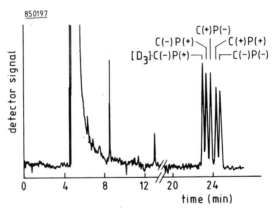

**Figure 7** GC separation of the stereoisomers of C($\pm$)P($\pm$)-soman and of $D_3$-C($-$)P ($+$)-soman on a type I Chirasil–Val/Carbowax 20M column. One side of the Carbowax column was connected with the injection port of the gas chromatograph. Injection of a 0.3-$\mu$l sample in isopropanol. Alkali flame ionization detection. See Materials and Methods for further details. Reprinted with permission from *Anal. Biochem.* **151**, 242 (1985).

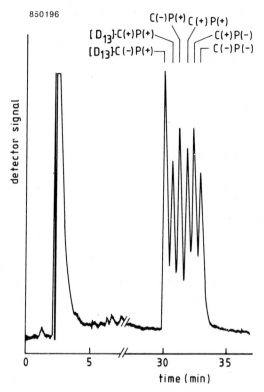

**Figure 8** GC separation of the stereosiomers of C($\pm$)P($\pm$)-soman and of the $D_{13}$-C($\pm$)P($+$)-epimers on a type II Chirasil–Val column. Injection of a 0.3-$\mu$l sample in ethyl acetate. Alkali flame ionization detection. See Materials and Methods for further details. Reprinted with permission from *Anal. Biochem.* **151**, 242 (1985).

Incubation of C($\pm$)P($\pm$)-soman in pretreated blood acidified to a final pH of 4.2 provided an adequate stabilization of the P($+$)-epimers for at least 40 min (Table II). However, the P($-$)-epimers levels were too high after incubation for 5 min and increased considerably during further incubation for 40 min. Our above-mentioned results suggest that this increase of concentration may be due to carboxylesterase reactivation induced by fluoride ions, although present at low concentration (about 10 $\mu$M), at the acidic pH. Accordingly, the regeneration is completely suppressed after complexation of fluoride ions with aluminum ions (0.2 mM) (Table III). However,

L. P. A. DE JONG *ET AL.*

**Table II**  Effect of incubation (pH 4.2, 0°C) in presaturated[a] rat blood/0.2 M acetate buffer on the concentration of added $C(\pm)P(\pm)$-soman (10 ng/ml blood)

| Isomer | Soman isomer[b] found as percentage of added amount after incubation for | |
|---|---|---|
| | 5 min | 45 min |
| $C(+)P(+)$ | $103 \pm 3$ | $101 \pm 2$ |
| $C(+)P(-)$ | $304 \pm 35$ | $1311 \pm 44$ |
| $C(-)P(+)$ | $98 \pm 1$ | $92 \pm 6$ |
| $C(-)P(-)$ | $252 \pm 26$ | $1141 \pm 73$ |

[a]By pretreatment of anesthesized, atropinized and artificially respirated rats with 6 LD50 of $C(\pm)P(\pm)$-soman 1 h before sampling.
[b]Means $\pm$ standard deviation ($n = 3$) corrected for initial $P(-)$-values. $P(+)$-isomers are not detected in presaturated rat blood.
Reprinted with permission from *Anal. Biochem.* **151**, 242 (1905).


**Table III**  Stability of added $C(\pm)P(\pm)$-soman (10 ng/ml blood) in presaturated[a] rat blood/0.2 M acetate buffer (pH 4.2, 0°C) in the presence of 2.5 mM aluminum ions, without and with added neopentyl sarin

| Isomer | Soman isomer[b] found as percentage of added amount | | | |
|---|---|---|---|---|
| | without neopentyl sarin after incubation for | | with neopentyl sarin after incubation for | |
| | 5 min | 45 min | 5 min | 45 min |
| $C(+)P(+)$ | $98 \pm 10$ | $106 \pm 4$ | $92 \pm 0$ | $93 \pm 0$ |
| $C(+)P(-)$ | $87 \pm 9$ | $56 \pm 36$ | $100 \pm 7$ | $102 \pm 6$ |
| $C(-)P(+)$ | $99 \pm 6$ | $95 \pm 3$ | $94 \pm 3$ | $98 \pm 6$ |
| $C(-)P(-)$ | $93 \pm 7$ | $54 \pm 36$ | $96 \pm 1$ | $98 \pm 1$ |

[a]By pretreatment of anesthesized, atropinized and artificially respirated rats with 1 LD50 of $C(\pm)P(\pm)$-soman 1 h before sampling.
[b]Means $\pm$ standard deviation ($n = 3$, without neopentyl sarin, or 2, with neopentyl sarin) corrected for initial $P(-)$-values. $P(+)$-isomers are not detected in presaturated rat blood.
Reprinted with permission from *Anal. Biochem.* **151**, 242 (1985).

a new problem arises. At these conditions, the P(−)-epimers partly disappear in blood samples taken from rats which have been treated with 1 LD50. Apparently, irreversible binding sites had not completely been occupied by this pretreatment for 1 h. We, therefore, acidified the blood samples with acetate buffer containing the soman analogue neopentyl sarin $(CH_3)_3CCH_2O(CH_3)P(O)F$, in a 100-fold molar excess relative to the added amount of $C(\pm)P(\pm)$-soman, in addition to aluminum ions. From the results of these experiments, as given in Table III, it is concluded that in this blood–buffer mixture the four stereoisomers of soman are satisfactorily stabilized for further work-up.[10]

Extraction of the soman stereoisomers from the blood–buffer mixture to which an appropriate amount of internal standard was added, is performed by adsorption in a Sep-Pak $C_{18}$ cartridge and subsequent elution with ethyl acetate. To increase sensitivity of the assay, eluates are concentrated at reduced pressure to a volume of about 0.1 ml. The overall recovery of the work-up procedure is about 50%[10] for each soman stereoisomer.

*Minimum detectable concentration* The described work-up procedure together with GC assay on a Chirasil–Val column (injection volume 0.3 μl) by using a NP-selective alkali flame detector allow the determination of soman concentrations down to ca 1 ng $C(\pm)P(\pm)$-soman/rat blood sample, i.e. ca 250 pg stereoisomer/rat blood sample (maximal 3 ml).

Samples obtained in toxicokinetic studies a relatively short time after injection of $C(\pm)P(\pm)$-soman, contain P(−)-epimers only which can be separately determined on a Carbowax or SE-54 column. By performing GC assay on a column with chemically bonded phase, larger sample volumes (up to 5 μl) can be analyzed with on-column or splitless (Grob) injection techniques. In this way the minimum detectable concentration of the P(−)-epimers is improved by one order of magnitude to 30 pg P(−)-isomer/rat blood sample.

*Toxicokinetics of soman stereoisomers in the rat* The described work-up and quantitative analysis method is now being used to follow blood levels of the stereoisomers of soman in the rat after $C(\pm)P(\pm)$-soman administration. In accordance with the *in vitro* results the P($+$)-epimers disappear rapidly from the blood. The concentration of the P($-$)-epimers can be determined up to 4 h after injection of 6 LD50 of C($\pm$)P($\pm$)-soman.

## Acknowledgement

The authors thank Matthijs Otto, George van den Berg, Ger Moes and Chris Schröder for isolation of soman stereoisomers and synthesis of deuterated soman analogues and Chirasil–Val, Carla Degenhardt for coating of GC columns and expert assistance in GC analyses, and Herma van der Wiel of the Medical Biological Laboratory TNO for performance of animal experiments. This research was supported in part by Research Grant DAMD 17-85-G-5004 from U.S. Army Medical Research and Development Command.

## References

1. O. L. Wolthuis, F. Berends and E. Meeter, *Fundam. Appl. Toxicol.* **1**, 183 (1981).
2. L. P. A. de Jong and G. Z. Wolring, *Biochem. Pharmacol.* **33**, 1119 (1984).
3. O. L. Wolthuis, H. P. Benschop and F. Berends, *Eur. J. Pharmacol.* **69**, 379 (1981).
4. J. H. Keijer and G. Z. Wolring, *Biochim. Biophys. Acta* **185**, 465 (1969).
5. A. J. J. Ooms and C. van Dijk, *Biochem. Pharmacol.* **15**, 1361 (1966).
6. A. J. J. Ooms and H. L. Boter, *Biochem. Pharmacol.* **14**, 1839 (1965).
7. P. J. Christen, Ph.D. Thesis, Leiden University (1967).
8. P. J. Christen, J. A. C. M. van den Muysenberg, *Biochim. Biophys. Acta* **110**, 217 (1965).
9. H. P. Benschop, C. A. G. Konings, J. van Genderen and L. P. A. de Jong, *Toxicol. Appl. Pharmacol.* **72**, 61 (1984).
10. H. P. Benschop, E. C. Bijleveld, M. F. Otto, C. E. A. M. Degenhardt, H. P. M. van Helden and L. P. A. de Jong, *Anal. Biochem.* **151**, 242 (1985).
11. E. Bayer and H. Frank, *Ger. Offen.* 3,005,024; *Chem. Abstr.* **95**, 151662e (1981).
12. E. Bayer and H. Frank, U.S. Patent 4,387,206 (1983).
13. E. Bayer and H. Frank, *ACS Symp. Ser.* 121 (*Modif. Polym.*) 341, Amer. Chem. Soc., Washington.
14. R. C. M. de Nijs, G. A. F. M. Rutten, J. J. Franken, R. P. M. Dooper and J. A. Rijks, in: *Proceedings, Third International Symposium on Capillary Chromatography* (R. E. Kaiser, ed.) 319, Inst. Chromatogr., Bad Dürkheim.
15. G. Schomburg and H. Husmann, *Chromatographia* **8**, 517 (1975).
16. H. P. Benschop, C. A. G. Konings and L. P. A. de Jong, *J. Am. Chem. Soc.* **103**, 4260 (1981).
17. L. P. A. de Jong and C. van Dijk, *Biochem. Pharmacol.* **33**, 663 (1984).
18. A. Verweij, E. Burghardt and A. W. Konings, *J. Chromatogr.* **54**, 151 (1971).
19. I. Abe, S. Kuramoto and S. Musha, *J. High Resolut. Chromatogr. Chromatogr. Commun.* **6**, 366 (1983).

20. P. J. Christen and E. M. Cohen, *Acta Physiol. Pharmacol. Neerl.* **15,** 36 (1969).
21. P. J. Christen, P. K. Schot and E. M. Cohen, *Acta Physiol. Pharmacol. Neerl.* **15,** 379 (1969).
22. L. Harris, C. Broomfield, N. Adams and D. Stitcher, *Proc. West. Pharmacol. Soc.* **27,** 315 (1984).
23. J. S. Little, C. A. Broomfield, L. J. Boucher and M. K. Fox-Talbot, *Fed. Proc.* **45,** 791, abstract 3703 (1986).

# Author Index

# Subject Index